Praise for CHASING CLAYOQUOT

A *Globe and Mail* Top 100 Book

"A Thoreau for Clayoquot ... Pitt-Brooke wins the reader over with the integrity of his voice ... it is the depth of his compassion and unwillingness to default to simple conclusions that bears the narrative into the reader's trust ... Like the best of Canada's naturalist authors, [he] possesses a lyric voice that rises distinctively from the Nearctic encounter with big landscapes, big water and big weather."

— *The Globe and Mail*

"Pitt-Brooke ... has penned a hymn to his surroundings. And, perhaps a prayer as well." — *Toronto Star*

"Pitt-Brooke has written a paean to ... a wondrous place."

— *Canadian Geographic*

"*Chasing Clayoquot* is wilderness veneration in the tradition of Thoreau, but Pitt-Brooke rewards the reader with beguiling facts of nature and history ... In the most moving passages, beyond 'the edge of the human domain,' Pitt-Brooke witnesses an alien world, 'out of my depth,' unnerved and awed by what he calls the 'scary and powerful' magic of nature. Chasing Clayoquot is education, inspiration and adventure in one of the world's most dramatic environments."

— *Dragonfly* magazine

"Pitt-Brooke's greatest talent is for teasing out the fascinating details of an ecosystem and then assembling them in a way that leaves you gobsmacked by the amazing interconnectedness of life."

— *Winnipeg Free Press*

"Poetic and heartfelt ... He makes the mosses, berries and organisms of decay as interesting as the whales and bears ... Pitt-Brooke finds not God but a sense of peace and meaning. And by telling us these stories he hopes to prompt our sense of wonder."

— *Victoria Times-Colonist*

"Pitt-Brooke is a most able guide, and *Chasing Clayoquot* is an admirable achievement — he's created a book that not only sings the praises of our area's natural surroundings, but is also praiseworthy in itself as a polished and elegant piece of contemporary nonfiction ... One third outdoorsman's personal history, one third naturalist ethics text and one third Linnaean guidebook, Pitt-Brooke's book is a wonderfully effective exploration of the natural processes of the sound ... Pitt-Brooke places himself in the tradition of A.Y. Jackson and the Group of Seven — the revelling in and revealing of the Canadian landscape for the Canadian people. And what they did in the last century in oils, he has succeeded in doing now with words." — *The Westerly News*

"Pitt-Brooke's thoughtful observation, scientific knowledge and love for his subject infuse his luminous prose ... *Chasing Clayoquot* is a book to read and then read again. And to think about for a long time"

— *Vernon Morning Star*

"With eloquence and a sense of joy, David Pitt-Brooke takes you there, in a way that will make you want to follow in his footsteps."

— artist and naturalist Robert Bateman

Chasing CLAYOQUOT

A WILDERNESS ALMANAC

Chasing CLAYOQUOT

David Pitt-Brooke

RAINCOAST BOOKS
Vancouver

Raincoast Books is a member of CANCOPY (Canadian Copyright Licensing
Agency). No part of this publication may be reproduced, stored in a retrieval
system or transmitted in any form or by any means without prior written
permission from the publisher, or, in case of photocopying or other reprographic
copying, a license from CANCOPY, One Yonge Street, Toronto, Ontario, M5E 1E5.

Raincoast Books acknowledges the ongoing financial support of the Government
of Canada through the Canada Council for the Arts and the Book Publishing
Industry Development Program (BPIDP); and the Government of British
Columbia through the BC Arts Council.

Text and cover design by Sari Naworynski
All photographs © Adrian Dorst 2004
Maps by Marjolein Visser

NATIONAL LIBRARY OF CANADA CATALOGUING IN PUBLICATION DATA

Pitt-Brooke, David
      Chasing Clayoquot : a wilderness almanac / David Pitt-Brooke. —
1st Canadian ed.

Includes index.
ISBN 1-55192-543-5 (bound.) — ISBN 1-55192-771-3 (pbk.)

      1. Clayoquot Sound Region (B.C.)—Description and travel. I. Title.
FC3845.C53P47 2004   971.1'2   C2003-906936-2

LIBRARY OF CONGRESS CONTROL NUMBER:

Raincoast Books   *In the United States:*
9050 Shaughnessy Street   Publishers Group West
Vancouver, British Columbia   1700 Fourth Street
Canada   V6P 6E5   Berkeley, California
www.raincoast.com   94710

At Raincoast Books we are committed to protecting the environment and to the
responsible use of natural resources. We are acting on this commitment by working
with suppliers and printers to phase out our use of paper produced from ancient
forest. This book is one step towards that goal. It is printed on 100% ancient-
forest-free paper (100% post-consumer recycled), processed chlorine- and acid-
free, and supplied by New Leaf Paper. It is printed with vegetable-based inks.
For further information, visit our website at www.raincoast.com. We are working
with Markets Initiative (www.oldgrowthfree.com) on this project.

Printed and bound in Canada by Friesens

10 9 8 7 6 5 4 3 2 1

To my parents Tena and Doug

Earth's crammed with heaven and every common bush afire with God
— Elizabeth Barrett Browning

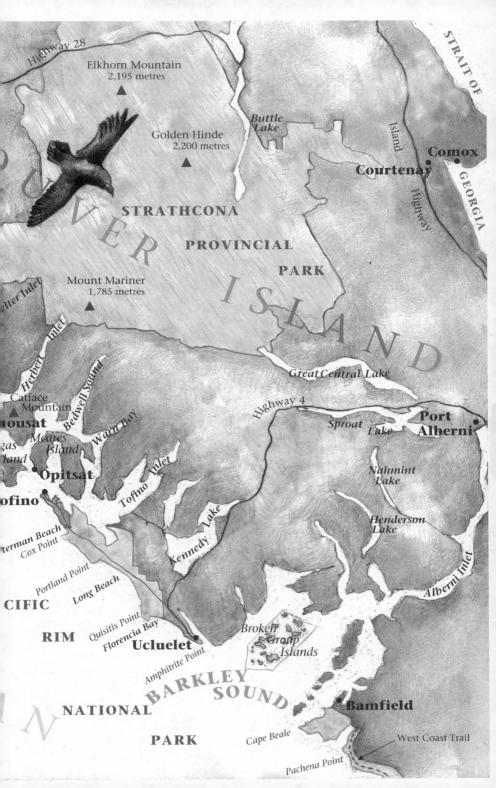

CONTENTS

FOREWORD

It is increasingly rare these days that citizens of the North American continent really know the place in which they live. Not that long ago, we lived off the land and our survival required that we be acutely aware of the natural world around us. Today, it is considered a luxury to have the time to watch for the signs of a season changing, to look for the return of a favorite song bird, or to take days "off" to hike into nature. David Pitt-Brooke has given us a bit of a reprieve. We can almost taste the salt air blowing off the roaring Pacific on Chesterman Beach. We can follow marbled murrelets as they fly over the ocean looking for a spot to roost in the old-growth forest of the Sound.

I am grateful for this book because it reminds me of what I loved best about being involved in the fight for Clayoqout Sound. I loved the place and the rugged people that call it home. I loved the fact that it was still there, still so forceful, still so beautiful and astounding in its diversity. From the first time that I visited in the early 1990s, I found a place that exceeded all my expectations. I was surrounded by snow-capped mountains that fed estuaries filled with fish and bird life. I could walk out on the mudflats at dawn to catch salmon and harvest mussels and oysters. I could trace the tracks of bears through the forests of giant cedar and Sitka spruce. It was a place that engaged all your senses and made you want to do what you could to ensure that it would never disappear.

Many of us fought for the Sound's future during that decade. We met on stormy nights to drink coffee with Native leaders and discuss their aspirations for their people and home. We attended countless meetings with government and industry officials all over Canada about the clearcut logging there. We took the battle home with us to Washington D.C. and New York to engage the public and private sector in shaping the Sound's future. One thing that never wavered throughout those years, one thing that continues to drive efforts for its conservation, is the pure life force of the Sound. That life force is beautifully captured in *Chasing Clayoquot* as David Pitt-Brooke takes us through the seasons while he embarks on his personal journeys out into nature, out into the Sound.

We are made better as a people by our alliances with places like Clayoquot Sound. With so much of nature being destroyed around us, we can take solace, as we might in the sanctuary of a church, in its wholeness and timelessness. While many of Mr. Pitt-Brooke's readers may never travel to the Sound, they will know something of what keeps drawing us back, of what makes us want to engage in bettering this miraculous natural world we have been so lucky to inherit from our ancestors.

ROBERT F. KENNEDY JR.

Prologue

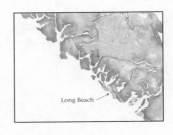

Long Beach

PROLOGUE: Beginnings

A bright May morning, the weather cool and clearing as I step from the forest onto the broad sweep of sand at the south end of Long Beach. Pausing briefly to admire the view and check my bearings, I move off in the general direction of Portland Point, 16 kilometres away to the northwest. Behind me is Quisitis Point, a mass of broken rock crowned with a dense growth of cedar and western hemlock. Between these two extremes of land lies the vast arc of Wickaninnish Bay with its cold-water breakers rolling in from the north Pacific, rank upon rank, endlessly.

My timing is perfect. The tide has ebbed far out into the bay, leaving wet sand agleam in the sunshine. Wisps of mist, coaxed upward by the growing warmth of the day, drift past, or rather, part to receive me as I move forward. The smooth even surface of the beach is broken, here and there, by dark reefs of bedrock, monolithic, slick with kelp and bejewelled with myriad multi-coloured shells. I am walking on the bottom of the ocean, temporarily relinquished, laid bare for my inspection by a parting of the waters. The air is pungent with the sharp odour of salt sea and it shimmers with sound: the rush of wind, the crying of seabirds and, above all, the throbbing, pounding, everlasting roar of surf.

To my right, dark forest encloses the beach, but to my left I can see clear to the ocean's horizon. Ahead, beyond Portland Point, the modest bulk of Radar Hill occupies the middle ground; farther north,

2

the peaks of Lone Cone and Catface Mountain rise abruptly from the surrounding countryside. Finally, faint in the misty distance, the long high ridge of Flores Island stands in outline against the blue. I walk and I walk, like a man chasing a mirage, never gaining on the distant horizon, mountain upon mountain receding through the haze of ocean humidity.

I was new to the west coast of British Columbia's Vancouver Island in those days, working as a biologist, naturalist and educator for the Canadian Parks Service at Pacific Rim National Park, at the southern edge of Clayoquot Sound.

I thought I'd gone to heaven. This is one of those rare places where the storehouse of nature is still full to the brim. The profusion and variety of life are a naturalist's dream. The job was a joy. I could take a group of people into the forest, stop anywhere, reach out and touch some marvel. I could escort visitors onto the beach, pick a tiny creature from a tide pool and drop it into a child's hand for closer inspection, then return the beast to its pool with no harm done to either. It was environmental education at its best and most vivid.

My companions on those little walks were a committed lot, passionate and eager to learn. The west coast of Vancouver Island is a good long way from anywhere, as the expression goes. Long Beach lies at the end of a twisting highway, hours of arduous driving. Nobody drops in, spur of the moment, for a quick look on their way to somewhere else. It takes motivation to brave the journey over the mountains. Visitors to the park wanted to be there. They recognized the value in wilderness experience. My kind of people.

My kind of country, too, and so it has remained. I am privileged to live in one of the world's most beautiful landscapes. Privileged and uneasy. Since World War II, the development of pristine wilderness has accelerated enormously. In recent years especially, wreckage and refuse have advanced into even the most remote areas. Places like Clayoquot Sound have become islands in an ocean of destructive development. Nor is this piece of countryside immune. The same fierce pressures are at work here; we're just a little behind the curve. As other beautiful wild places are lost, one after another, my uneasiness grows. One of these days it might be our turn.

That is reason enough to undertake the grand tour of the sound and its adjacent countryside that I'm proposing here. Call it a rediscovery. I shall go to enjoy the loveliness, of course — I hope to reawaken something of the sense of wonder I felt on that first day on Long Beach. But I go also to reassure myself that the beauty is still there, that the worst has not happened. I want to pay my respects and to store up remembrances against the possibility — perish the thought — that these treasures could yet be taken.

But I have other good reasons, many thousands of them.

This place with the odd name, Clayoquot Sound, a land- and seascape of dark forests, steep mountain slopes and fractured islands along Canada's exposed west coast, exercises an extraordinary enchantment, a powerful charisma. Visitors come from all over the world, hundreds of thousands of people annually, drawn by the area's reputation for unspoiled natural beauty or by its notoriety as an ecological flashpoint, a symbol of embattled wilderness.

They come with an urgency that sometimes borders on desperation, these refugees, searching for wonderment and beauty in the natural world, qualities rare in their ordinary daily lives. They come for understanding and enlightenment. They come to experience an alternative lifestyle, seeking some sense of acceptance in this place, if only in a modest capacity as observers. They come to touch the earth, looking for an antidote to their growing isolation from the natural world, the bane of modern urban existence.

But few have the time, the means or the information needed to travel much off the beaten path. They wander the village streets. They shop the galleries. They sift through all the commercial hype that plagues the modern tourist, hoping for some grain of insight. After a few days they head home, relaxed and pleased with the beauty of the scenery, but with only the most superficial awareness of this very special place. Many must depart feeling vaguely cheated, denied the fullness of an experience they've come so far to embrace. And what about all those others who feel the same call to wild and beautiful country, but can only dream of travelling to places like Clayoquot Sound? How can they find satsifaction?

I think I can help.

Let us spend a year getting the measure of this place through all its changing seasons. We will venture out at least once in each of twelve calendar months. Each journey will focus on the events and attractions particular to its season: the special places, notable species of plants and animals. With a little luck, there will be glimpses of nature's true face — an elusive prize even in Clayoquot Sound — moments of vivid beauty revealed in different moods and manifestations: the tumult of a winter storm, bright spring mornings, warm still evenings in sheltered valleys where the view hasn't changed substantially in five thousand years. By the time we're finished we'll have re-acquainted ourselves, at least in passing, with much of what is unique and extraordinary in this unique and extraordinary place.

And if we pay proper attention, we may even be able to capture something of the strength and power that dwells in true wilderness. In the old times, a novice shaman or anyone else seeking mystical power could undertake a spirit quest, pursuing a personal encounter with the supernatural. Spirits were everywhere in wild country. The seeker might encounter the *Ya'ai*, a man-like creature with tufts of feathers growing from each side of its head. *Ya'ai* could bestow shamanistic powers, wealth, success in whaling. Or the seeker might come upon a supernatural animal — squirrel, mink or raven — shaking a tiny rattle and singing over a groaning, writhing log. Wolves were a particularly potent source of power. The properly prepared and purified seeker, shaking off the intoxicating effect of the spirit's presence, uttered a ritual cry to dominate and control the apparition: *"'Ai."* At the seeker's cry, the vision would vanish, leaving some object in which the power of the encounter resided. Later the spirit would return to the seeker in dreams, to teach the ritual displays, dances and songs of power, the tangible signs and symbols of the encounter. It was a serious business, a dangerous business, this pursuit of supernatural powers. Spirits could destroy the poorly prepared seeker.

I don't suppose we'll actually encounter *Ya'ai*. But I've always thought that if we approach wild country in the proper frame of mind, well prepared, it can bestow not only the gift of understanding, but also some sort of expansion to the spirit. A lifting up, a strengthening, an enthusiasm. And later, perhaps, it can teach us to sing the songs of power.

January

WIYAQHAML*

*(No Food Getting for a Long Time Moon or
Not Outdoors Moon)*

*in the language of the Nuu-chah-nulth people

January is fierce and dark on the west side of Vancouver Island, the windward shores. One storm after another ploughs across the coast, bringing heavy overcast, blustery wind and torrential rain. Freezing temperatures occur only occasionally at sea level, but conditions are more severe in the mountains. Tremendous depths of snow accumulate.

In the old times, deep winter must have been a season of terrible anxiety. Stores of food ran low. If the fall harvest of salmon had been meagre, or if the weather was especially severe, starvation was a real possibility. The people were utterly dependent on the natural world for their supplies, bound to the rhythm and march of the seasons. Men went fishing for cod, when seas were not too rough, or set traps for bait: kelpfish and perch. Tribes wintering further inland hunted land mammals, especially elk and deer wallowing through deep forest snow. Women took advantage of low ebb tides to gather clams and other marine invertebrates, or to fill baskets with the last of the evergreen huckleberries. In hard times, people might be driven to collect small mussels and other less desirable shellfish, or, in cases of dire necessity, to comb the beaches for storm-killed carrion.

Even now, January is an anxious, waiting time. The holiday festivities are over; the children are back in school. People retreat indoors, waiting for easier weather, feeding the fires, watching for leaks and drips whenever the southeasterly gale blows. Confinement and darkness take their toll. The wintertime blues are a fact of life for west coast people. The streets are empty, business is slow. A fortunate few, much envied by their neighbours, escape to tropical climates for a few days of sunshine and warmth. The weather is too stormy for most outdoor work. Crab fishermen watch for breaks in the storm to tend their pots. Hydro and telephone linemen, exceptions to the rule, ply their trade busily at all hours of the day and night, repairing lines brought down by falling trees.

CHASING THE TEMPEST: Who Has Seen the Wind?

A developing low, well offshore, will rapidly deepen to become a 966-millibar centre, 90 miles southwest of Cape St. James. The associated frontal system will cross Vancouver Island late Thursday afternoon. Seas will rise to 10 to 13 metres along the outer coast.
— Environment Canada, Marine Weather Forecasts

I stand in night-time shadow at the bottom end of First Street in Tofino, where the worldwide network of pavement ends at the Pacific Ocean. Half a dozen sodium vapour lamps illuminate the heavy, rain-slick wooden planking of Government Wharf. Through each cone of yellow light, the deluge falls in wavering sheets, like something out of a Hollywood movie. There might be a stage crew with a firehose hidden somewhere in the darkness above.

The gleam from the lamps reflects off the road and off the dark, rippled water below the pier. It lights the offices of Tofino Air, long since closed for the day, and the parking lot of the Sea Shanty restaurant, long since closed for the season. The streets running up the hill behind me are utterly deserted. The village seems empty of people on this wild night.

Rain beats down with such force that droplets of water rebound from the pavement, trying to regain the air. A mist of fine spray hangs over the asphalt. Water pools and runs in rivers down the steep incline of the street, through the parking lot, over the curb and across the narrow cobble beach into the ocean.

Fierce gusts sweep down the hill, jostle past, race off across the parking lot and out along the pier. To my left, the sign affirming TOFINO, BC — PACIFIC TERMINUS TRANS-CANADA HIGHWAY seems solid enough, but over at Tofino Air, the sign advertising SCENIC FLIGHTS wobbles and creaks in the turbulence. On one side of the dock, three floatplanes rock in the choppy water. On the other side, two fishing boats pass a similarly restless night. The clang and clatter of rigging, gear and aluminum masts carries through the darkness over the water. The world beyond the farthest reach of the pier is blackness, a void.

The west coast of Vancouver Island is one of those places, relics of another age, where people live in defiance of wind and weather. The latest forecast from Environment Canada or Coast Guard Radio is a topic of real interest, a matter of life and death. Even in an age of radar, radio and satellite positioning, the gales still come out of nowhere, boats go down and mariners die.

We get more than our fair share of heavy rain, fierce winds and rough seas, especially in winter. Statistics tell the story. Measurable precipitation occurs on 202 days of the year. The airport weather station averages 3.3 metres of annual precipitation, mostly in the form of rain or drizzle, with a maximum one-day record of 18.4 centimetres. (By way of comparison, Vancouver receives just over one metre per annum, a third the Tofino total, while Los Angeles records a paltry 46 centimetres.)

Out on the coast we can at least take some consolation in knowing that conditions are even more severe among the mountains of the island's interior, where the most intense precipitation occurs. The Canadian record for annual precipitation is held by the weather station at Henderson Lake (947.9 centimetres in 1997) and the record for daily precipitation is held by the weather station at the now-defunct Brynnor Mine near Maggie Lake (48.9 centimetres on October 6, 1967). Think about that, half a metre of water in 24 hours, laid down across the entire landscape — lakes, rivers, valleys, mountains — and every drop of it striving to get back to mother ocean by the quickest possible route.

On the other hand, when it comes to temperature, the west coast of Vancouver Island may have Canada's mildest winter climate. Snow and freezing temperatures are uncommon at sea level. Again, numbers tell the tale. The thermometer at Tofino airport rises above freezing an impressive 364 days a year, and daily minimums for January average a balmy 1.2°C. Unfortunately our summers, though relatively dry, aren't an awful lot warmer than our winters. It's the proximity of the north Pacific that does it. Trying to get a tan on Long Beach is like sunbathing in front of an open refrigerator.

West coast people have a love–hate relationship with their rigorous environment. On the one hand, they fear the elements, the turbulent heart and darkest side of Clayoquot Sound. Not for nothing do all those people depart hastily in search of tropical sunshine when the first autumn storms roll in. On the other hand, that same rawness, the elemental flavour of Clayoquot Sound, has an undeniable appeal. It's clean and uncomplicated. It's authentic and thrilling. Not for nothing do all those people make the treacherous journey west through the mountains, hoping to see a storm on the outer coast.

Those same physical elements profoundly affect the natural world. A given set of conditions fosters a characteristic group of species, gifted through accident and evolution with the right equipment for survival and prosperity. Other species fail to thrive because they are somehow lacking. There are other factors in the equation, including luck, and much depends on interactions between different individuals and species, that mix of competition and synergy. But the physical environment is fundamental in shaping the biological community.

Just as desert communities are shaped and defined by heat and lack of moisture, and Arctic communities by bitter cold and blowing snow, natural ecosystems in Clayoquot Sound are the product of year-round mild temperatures, rugged topography, the proximity of a vast, cold, restless ocean and, most especially, wet, tumultuous winter weather.

The air is quiet when I step from the forest onto Chesterman Beach, a smaller version of Long Beach some four or five kilometres south of Tofino. But the western sky is one wall of cloud, so dark as to be almost purple, and nearly featureless except for faint, far-off curtains of rain, like a watercolour wash. So far, so good; it's a promising start to my storm-watching expedition. The horizon, where dark cloud meets even darker ocean, is a clear, straight line, sharp as a razor.

The tide is extremely low. The Pacific has retreated far out into Cox Bay. Frank Island, a small, rocky mass capped with redcedar and spruce, perhaps half a kilometre offshore, is temporarily connected to the beach via a natural causeway of sand: a tombolo. From my low angle of perspective, the dark background lends a looming quality to the little island, an unnatural clarity. It seems to stand at the edge of creation, the exposed bony rim of the world just a few hundred metres away.

The dark cloud arches high overhead, a vaulted roof extending east and north, gradually overtaking patches of brighter sky beyond Catface Mountain and Flores Island. The sand is flat as pavement, grainy, dun coloured, packed damp but not wet.

I start across the beach, heading for Frank Island. Away from the trees I can feel the lightest breath of wind flowing from the south. On Cox Bay, south of the tombolo, the breeze has rubbed the surface of the ocean into tiny ripples, lending the water a ground-glass appearance, a slatey, grey-green reflection of the looming cloud beyond. On the north side of the tombolo, the sheltered waters still have a molten sheen, rippling blue and gold, reflecting lighter sky to the north.

Five minutes of walking takes me to the island. At closer range, the rocks are jagged and raw. The outside of the island is exposed to the heaving waters of the open Pacific. These are not choppy little waves stirred up by a local wind but great ocean swells spawned in some distant storm, heaven knows where on the broad expanse to the west, now sedate and rounded with travel.

As each swell reaches the island, the water along the shore lifts, boils and breaks into foam. Each rock and pinnacle wears a collar of purest white. The swell subsides, its force spent, seawater runs off in rivulets and the ocean retreats, gathering itself to launch the next wave. The air is full with the muttering voices of the surf, the slap of waves, the gurgle and suck of seawater in the cracks and crevices of the rocks and surge channels that dissect the island.

I climb a nearby pinnacle. Now the horizon is blurred and indistinct, smudged with rain. To the north, a dark curtain falls across the bright sky, blotting out the blue. The islands in the middle distance fade to grey.

Away to the west, across the deep waters of Templar Channel, is Lennard Island, not yet obscured by the rain, home to the Lennard Island light station: a cluster of buildings, white walls, red roofs, a single window shining with interior light. The beacon in the tower sends two great beams of light in opposite directions. Visible in the thickening air, each beam traces out half of a giant circle, painting the length of Chesterman Beach every eleven seconds.

The leading edge of the squall blows in as I leave the island. I feel a puff of wind at my back and the first sprinkle of rain. From this distance the forest growing on the low ground behind the beach, tall Sitka spruce and western redcedar, has the look of an enormous hedge, punctuated here and there by the twinkle of electric lights. It is too early for a general lighting up, but when real darkness comes the myriad bright windows — mine among them — betray the many houses that have crowded along the beach in recent years. For now, the premature twilight preserves an illusion of wildness. The seaside mountains of Clayoquot Sound rise in a sweeping semicircle from southeast to northwest. The·great peaks of central Vancouver Island are white with snow and half obscured by cloud.

By the time I reach the shelter of the forest, Cox Bay is covered in choppy whitecaps. Rain patters upon my shoulders. But when I turn for a better look, there is already a bright line along the horizon and a hint of sunlight beyond the cloud. For all its fierce appearance, the squall is passing.

By the time I reach the parking lot, the foghorn on Lennard Island is announcing in basso profundo that the air over the channel, now thoroughly humidified by drizzle, has condensed into fog. *Beware*, the horn bellows, separating the two syllables. *Be-Ware*.

The Earth's atmosphere is densely concentrated at sea level and thins rapidly with altitude. Three-quarters of all air molecules, by mass, are within ten to twelve kilometres of the surface. Half are within five or six kilometres.

Lone Cone Mountain rises steeply from the ocean five and a half kilometres north of Tofino. Catface Mountain is twice as far away, about eleven kilometres. If you are standing on the First Street dock, you can see both mountains. Now project those same distances straight up. Between you and Lone Cone is one-half the Earth's atmosphere. Between Lone Cone and Catface, another quarter of the total. Beyond Catface, to infinity, the remaining air molecules — the last quarter of the Earth's atmosphere — are so thinly scattered as to be almost inconsequential, as far as human biological functions go, anyway. There lies the lung-sucking emptiness of space and rapid death from oxygen starvation, a mere eleven or twelve kilometres away. It is that close. This air-ocean, the atmosphere on which all our lives depend, is perilously shallow. You might think we'd take better care of it.

The first nine to sixteen kilometres of atmosphere above the Earth's surface is termed the troposphere. Most of the turbulent atmospheric events we call weather — moment-to-moment changes in cloud cover, air temperature, wind, air pressure, humidity, precipitation and so on — occur in the troposphere. At greater altitudes the air is extremely rarefied, and vertical movement is largely absent.

So when you look at satellite images of the weather — fronts and systems and spiralling clouds — it's all taking place in an extremely thin film over the surface of the planet. Think of the layer of varnish on a wooden table. Now imagine weather, storms, all that ferocity, taking place *inside* the layer of varnish.

Back in my study, I unearth a thick sheaf of mimeographed typescript, faded with age. The Lennard Island light station went into service in 1904 and the first keeper was a young man named Frank Garrard who had emigrated from England in 1888. He worked for a time in Nanaimo and Port Alberni, then moved to Tofino, where he and his family lived for many years. Eventually he retired to Victoria and spent the last years of his life organizing his diaries into the monumental autobiography that I now hold in my hands — 350 pages in all, an extraordinary glimpse of daily activities on the outer coast during the early years of this century.

It was a hard life. Garrard and his contemporaries lived very close to the elements without the insulating effect of modern technology. The lightkeeper's accounts are full of close calls and harrowing encounters with wind, storm and wild seas, all in a day's work:

The men finished work on the fog station before the end of December 1905, and intended leaving the day before Christmas ... But on December the 23rd and 24th, a heavy gale got up making it impossible to get on or off the Island.

On the night of the 24th the gale reached its culmination and for a time was blowing a hurricane. It was during this that a barque, the Pass of Melford, *went ashore close to Ucluelet harbour and was lost with all hands, nothing but the upper part of some of the masts showing the following day. On Christmas, we spent the day watching the huge seas breaking over the line of reefs which protected the island to the westward. The reefs were about 50 feet high and extended across to the west of the island, forming a channel. At one moment the*

rocks would be standing up against the skyline, then
the next moment one of the big westerly swells formed by
the gale of the night before, would overwhelm the reef so
that nothing but a mass of broken water and spume
would show. Then the reefs would again appear, but a
cataract of water would be foaming down over them.

This went on all day and it was not until the evening it
moderated. Meanwhile our boarders spent the evening
with us and we had an evening's entertainment of games
— using the old fashioned ones of Family-coach; Earth,
air, fire and water; etc. I do not think anybody was able
to get ashore before the 26th of December, as the swell
was very bad for some days.

— Frank Garrard, *Memoirs*

The visit to Chesterman Beach and Garrard's autobiography have got me thinking about the possibility of a visit to the Lennard Island light station. It's hard to imagine a more spectacular location for weather watching. Lennard Island sits on the ragged edge of Clayoquot Sound, the last bit of rock before Australia. I wonder if visitors are welcome. When I get lightkeeper Andrew Findlay on the phone he turns out to be friendly and forthright. "Sure," he says. "Why not? Come on out."

There remains one slight problem: getting out there — and back again. Findlay suggests a call to the local Coast Guard Rescue Station. "They often make trips to Lennard Island as a courtesy," he tells me, "bring me the mail, spare parts and so on. Perhaps you could hitch a ride."

The morning is grey with a low overcast and soaking drizzle. The helicopter perched on the dock wears coast-guard colours: fire-engine red with a broad white stripe down the flank. The pilot is Simon Lebel, a flamboyant bearded Québécois.

I compliment Monsieur Lebel on his aircraft, which turns out to be a Messerschmitt MB BO-105. "Yes," he says, "a fine machine, by far the most manoeuvrable helicopter I've flown." And tough, evidently, though a little temperamental — as if Ferrari had manufactured a pick-up truck. "You must fly it the whole time," he adds.

Today he's moving people, ferrying coast-guard officials between stations. In fact, he's just leaving for Lennard Island to pick up passengers for the trip to Estevan light station, thirty kilometres to the north. He can drop me at Lennard now and collect me on the return trip.

Of course, there is always a possibility that deteriorating weather conditions or some other emergency will come along to change those plans. In that case I could be spending quite a bit more time on Lennard Island than I'd bargained for. Ah, well, life is full of uncertainties. Damn the torpedoes. Full speed ahead.

Minutes later, I am extensively and firmly belted into the left front seat of the chopper. Lebel is in the pilot's seat to my right. Drizzle beads up on the canopy and runs off in streaks. The whine of turbines rises to a shrill pitch. When Lebel introduces fuel, I feel the thud of ignition through the seat and a moment later catch a whiff of hot jet exhaust. The low-revolution alarm shrills, ominous sound, until the engines reach full speed. The rotor is whirling.

Lebel informs harbour control, then eases up on the collective control with his left hand, increasing the rotor pitch, and works the stick with his right. The aircraft quivers and trembles like a living thing. Suddenly the ground subsides beneath us, just a couple of feet, leaving the helicopter hanging from its rotor, swaying a trifle. Then we're rising fast, so quickly I scarcely notice the transformation as Tofino dwindles into miniature below us. One moment it's the usual scale and the next instant it's a toy town.

The height also lends a new perspective on the landscape. What had seemed solid ground, mountain upon mountain, reveals itself from an aerial view to be fragments separated by a lacework of water. I glimpse the true nature of Clayoquot Sound: an immense embayment of the Pacific Ocean crammed with islands, large and small. The names on the chart are freighted with history: Indian Island, Meares Island, Wickaninnish Island, Vargas Island, Flores Island, Father Charles Channel, Browning Passage, Lemmens Inlet.

The helicopter follows Duffin Passage to seaward, passing between Felice Island and Grice Point. Then we bear to the left, following the channel southward. Lennard Island is already visible in the middle distance, just 2.5 nautical miles (4.6 kilometres) away, hardly a hop in the helicopter.

On the right, beyond an archipelago of barrier islands — Wickaninnish, Echachis and Tonquin — lies the open Pacific. The altitude has extended my view considerably and still there is no end in sight to all that water. From the beach, the horizon is less than 10 nautical miles (18.5 kilometres) away; from 300 metres up, it is as far as 40 nautical miles (74 kilometres) and the ocean looks immense. But Australia is still over 6,000 nautical miles (11,100 kilometres) beyond the horizon. What I'm seeing right now is barely the edge of the edge.

Up here, we're conspicuously cocooned in technology, warm and dry. I look at the water far below. Our height has reduced the southwest swell to trivial proportions, small corrugations crawling across the surface of the sea.

Then I notice surf breaking across a set of submerged reefs in the channel between Tonquin and Echachis. It occurs to me that the swell must actually be quite severe; in calmer weather that passage is unruffled. It also comes to me that I'm only able to grasp this because I've been down there in rough weather and seen it for myself at close range. From this serene height, without that experience, I wouldn't have a clue. It's an important lesson. If we really want to know the natural world we have to get up close and personal, close enough to touch, close enough to engage all our senses.

Lebel approaches the light station from the east, banking around the south end of the island, losing altitude. We come in low past the tower and set down on a grassy meadow below the station. I thank him for the ride. He promises to give me a call on the radio when he's ready to head back to Tofino.

Free of the harness, I ease out of the helicopter and move away, crouching to avoid the whirling blades. Halfway up the path, I pass the two officials, bright with rain gear, hurrying to catch their ride. The lightkeeper is waiting to greet me. We are just shaking hands when a rising whine from below signals the Messerschmitt's departure. The helicopter lifts, banks and heads north, passing from view beyond the trees. The relative quiet, as the sound of the machine fades away, is remarkable.

The Lennard Island light station occupies a slight promontory on the southwest side of the island. On the north side of the tower are two cottages for the keepers and their families. A separate building, just south of the tower, houses the foghorn and engine room.

The drizzle has stopped, the clouds are brightening and the sun is threatening to break through. The swell breaking on the reef beyond the tower is impressive, but the wind has dropped to the merest breeze. If I was hoping to see fierce weather out here, I'm out of luck. Findlay consoles me with the grand tour: foghorn, engine room, tower. The whole facility is spotlessly clean and freshly painted. Without such zealous maintenance, these highly exposed buildings would deteriorate quickly. The spit and polish reminds me of a military installation. There is that flavour of military readiness with gear stowed, decks cleared, equipment locked and loaded. It's a routine state of yellow alert against an ever-present enemy: wind, weather and salt sea spray.

While we wait for the helicopter to return, Findlay invites me into his home for a warm-up of hot coffee and biscuits. The place is snug, warm and pleasant. He is relatively new to Lennard Island, having spent most of his career in stations farther north. It is not an easy life. Keepers and their families must contend with isolation, long periods of close confinement, personality clashes, hard, dangerous physical labour and the weather. Always the weather. It's not so very different from Garrard's day.

At four o'clock, true to his word, Lebel radios to say that the Messerschmitt is on its way. Findlay walks me down to the helicopter pad. As we wait, he points out the remains of a concrete boat ramp built in Garrard's time, now badly battered. The boathouse itself is gone, taken away by some long-forgotten winter hurricane.

Usually it's the faint, far-off popping of blades that announces an inbound helicopter. But this time I see the Messerschmitt before I hear it. From somewhere high inside the mist above McKay Reef on the seaward side of Wickaninnish Island comes a twinkle of landing lights. For just an instant I'm treated to an unearthly vision, a tiny star growing in the clouds. Then I hear the blades. The machine comes sliding down out of the sky, straight in, hovers tumultuously and settles onto the pad. I squeeze into one of the rear seats. It's crowded in there with me and my pack plus the officials and all their gear. No room for Garrard. Not his style, anyway; he comes and goes by boat:

> Once, when Lilly [Garrard's eldest daughter] and I had gone in [to Tofino], a heavy gale got up from the southeast. I did not wish to miss getting back to the lighthouse for that night, so when the tide had turned and was running out, we started, intending to use the strong ebb tide to help us row against the wind and sea.
>
> But as we got abreast of Village Island we found that the ebb tide against the strong wind had got up a terrible rip, especially as the wind had about reached its peak. Between the seas we made a little head-way, rowing, but as soon as we mounted the top of each wave, not only did we get deluged with spray, but for the moment came to a standstill. The tide was taking us out,

so by continuing rowing we gradually worked our way
through this rip.

I remember talking to Lilly about something we might
do "when" we got out to the island, but Lilly corrected
me, and said "if" we get out to the island.

It turned out not so bad as that. Just after we had
passed Village Island, the wind moderated and a deluge
of rain came on, which had the effect of deadening the
sea. By pulling hard to take advantage of the spell, we
made the island all right, and were greeted by [assistant
keeper] Pollock telling us that we were, or rather that I
was, a blamed fool for trying to come out at all in such
a gale.

Moments later we're back on the coast-guard dock. The light is
failing as we pull our gear from the helicopter. I say thank you and
goodbye. It was a most interesting experience, though I can't help
feeling a little disappointed. It's not that I'm ungrateful for the
smooth passage, no indeed, but the tempest has eluded me again.
Another false start. But as I climb the path to the street, I feel the first
faint gust of wind from the southeast. My time is coming.

A little meteorology, perhaps.

Picture an enormous river, a turbulent river, with vast tributary
streams rushing in from either side. Picture the glint of sunlight from
the pressure lines, swirls and eddies. The circulation of water in a
river is a useful analogy for thinking about the flow of air in the
atmosphere. The west coast of Vancouver Island lies in the path of a
vast river of air flowing across the Pacific. Warm air masses from the
south and cold air masses from the north pour into the main stream,
generating enormous eddies in the atmosphere.

All that atmospheric turbulence is driven by variations in solar
heating across the surface of the globe.

In the tropics, daytime heating is intense. Warm, humid air boils upward into the atmosphere. Water vapour condenses, heating the air still further. Tremendous storm cells develop. Turbulent air spills toward the subtropics, subsiding toward the surface. Some circulates back toward the equator. Some continues to spill northward into the temperate zones — as outbreaks of relatively warm, humid, high-pressure subtropical air.

Over the poles the situation is very different, especially during winter. Solar radiation is minimal and temperatures are extremely low. There is no liquid water; the air is very dry. The atmosphere is cooled, rather than heated, by the Earth's surface. With nothing to warm it from below, polar air is very stable. Great masses of exquisitely cold air accumulate over the wintertime poles. Being so cold, Arctic air (to talk of the northern hemisphere) is very dense, and pressure at the surface is high. Periodically, portions of this dense mass break off and avalanche toward the temperate zones — as outbreaks of cold, dry, high-pressure Arctic air.

Where Arctic and subtropical air masses meet and mix in the temperate latitudes, they create a zone of characteristically turbulent, stormy weather. The trough of relatively low-pressure air between converging warm and cold air masses is often referred to simply as a *low*, as in "a deepening low off the west coast will move rapidly inland overnight." Both warm and cold air masses end up orbiting the low in a counterclockwise direction. This complex system — a low-pressure trough surrounded by converging air masses — is termed a *low-pressure system* or *storm system*.

The area of contact, the partition, between different air masses is termed a *front* (I picture the billowing membrane that separates two enormous soap bubbles, such as children make on a summer's day). Such partitions may be hundreds of kilometres long and thousands of metres high.

The shape of each front depends on the nature of adjacent air masses. A warm subtropical air mass is less dense than a cold Arctic air mass at the same pressure. When a subtropical air mass shoulders into a slower-moving Arctic air mass, the leading edge of the warm air leaves the ground to flow up across the cold air.

This is a *warm* front, and it takes the shape of a long, shallow ramp. The thin leading edge of the warm air mass can extend a hundred and fifty kilometres or more over the undulating upper surface of the cold air mass. As subtropical air rises, it cools. Water vapour condenses into stratus cloud, thin at first but thickening to drizzle, then steady rain.

The leading edge of a vigorous Arctic air mass — where it circles the west side of a low and ploughs into the flank of a subtropical air mass — is called a *cold* front. A cold front is more abrupt than a warm front and often faster moving. Dense Arctic air moves along the surface, bulldozing its way under retreating subtropical air. The warmer, less dense air is lifted and cooled rapidly, with sudden condensation of water vapour. Cold fronts are turbulent, dramatic, with heavy intermittent rain squalls from towering cumulonimbus clouds.

On Chesterman Beach a rising breeze from the southeast signals an approaching system. The sky is hazy with high overcast, gradually thickening. This is the thin leading edge of a warm front. As hours pass, the overcast deepens. A light drizzle begins to fall. The southeast breeze becomes a wind, then a gale with steady rain. The storm rises to a crescendo as the trailing sea-level edge of the warm front passes. With the arrival of the warm air mass, the wind veers to southwest, temperatures rise, rain eases off to spotty showers and the day brightens. The "pineapple express" has arrived; you can almost smell the tropical flowers. Sometimes the sun shines wanly through the stratus clouds. Otherwise, drizzle and low overcast persist.

Either way, it doesn't last. The stratus deepens. Drizzle gives way to rain. The wind rises, veering farther west. Stratus overcast gives way to billowing cumulonimbus as the cold front arrives. The precipitation changes character, becoming fierce and squally with buffeting winds: one moment pounding rain, big drops, maybe even some hail; the next moment easing away to nothing; then pounding down again. From below, the clouds are dark and threatening, a heavy overcast. There may even be some thunder.

With passage of the cold front, the wind veers to the northwest. The overcast breaks up into puffy cumulous clouds. The air is clear, cold and unstable. Squall follows squall across the beach, but there are patches of blue sky, even some bright sunshine.

By now the larger part of the storm system has passed to the east. If there is to be a period of good weather, the winds will settle in the northwest and ease to light and variable. Sooner or later they veer back to the southeast as the next system approaches.

In the northern hemisphere, the zone of contact between Arctic and subtropical air masses shifts north or south depending on the season. In summer, Arctic outbreaks become less frequent and less intense. The tropics shift northward. The zone of contact between Arctic and subtropical air masses not only shifts into higher latitudes but becomes less turbulent. Fewer storms arise. Those that do are less energetic and they tend to follow a path that carries them well to the north of Vancouver Island.

Summertime finds us, generally, deep inside a subtropical air mass, warm, humid and stable. The wind blows moderately from the west or northwest. Daytime heating on land creates a chimney effect, especially in the mountains. Warm air rises, drawing a brisk after-noon breeze of cooler air off the ocean. Skies are usually clear. But Chesterman Beach may not enjoy much sunshine. Warm, humid air overlying the cold ocean condenses into fog or low marine cloud that drifts across the beach in the onshore breeze, to dissipate inland over sun-warmed forest.

In winter the situation is reversed. The tropics shift southward. Arctic outbreaks become increasingly frequent and energetic. The turbulent zone of mixing moves south and Vancouver Island finds itself once again on the storm track, catching one system after another.

Another late afternoon at Chesterman Beach, and this time the air is anything but quiet. I hear the moaning of the wind and the roar of surf long before I reach the edge of the forest. Trees creak and bend. Branches thrash. When I step into the open, the gale claws at my clothing, stings tears into my eyes, takes my breath away. Rain drives into my face, seeks out the chinks in my vinyl armour and trickles down my neck.

There is no sky, no cloud, no ceiling. There is only a pale nothingness above the darker nothingness of the land. Vision reaches out a certain distance and fails, quenched by the downpour. Cox Bay, Frank Island, Chesterman Beach, Lennard Island are at the limit of perception, as blurred as phantoms.

Water is everywhere. The air is full of it. The land is drowning in it. The entire sandy surface of the beach is covered by a liquid sheet draining toward the ocean. Raindrops pock and spatter. The wind raises wavering ripples, pushing them uphill against the current.

The surf zone near the beach is a wilderness of white water, rank upon rank of tumbling waves, the crash of individual breakers lost in the general din. The air throbs with one prolonged surging roar, a tremendous noise. Farther out, the whole vast surface of Cox Bay undulates, slopping hugely like water in a bucket. House-sized waves grind, like implements of war, steadily toward the land.

The rocks around Frank Island are slick with the wet. I drop into the lee of an immense half-buried boulder to catch my breath. The sudden stillness, the relative quiet in the calm pocket next to the boulder, is astonishing. The wind flowing around the rock is so strong it snatches sound away. The waves breaking before me are oddly muted, as if someone had turned down the volume.

Climbing to the view, I slip and gouge my hand. Blood streams away in the rain. The open ocean is a vast field of breaking waves, enormous, wind-driven. Where the surface of the water is visible, it has lost its normal lustre completely, become an unnaturally flat green or slate grey. But the overall impression is of whiteness. The sea is dominated by blossoming whitecaps. The wind, impatient with their progress, shears off the foaming peaks and carries the spray across the surface of the water or into the air. Suddenly I'm thinking of snow, white snow blowing across the dark of an airport tarmac.

A breaking wave in deep water moves with dignity, slow and ponderous. In sunlight, it's a beautiful thing to behold, a mountain of green water, its shoulders cloaked in a mantle of snowy white foam, massive, sedate, majestic. In the dark of a storm, the same wave becomes ominous and menacing. These are Kipling's "great, grey-green, hissing widow makers."

That sense of ponderous dignity is deceptive. As with elephants, size creates a slow-motion illusion. When great waves approach the shore their real speed becomes apparent. In shoal water off the mouth of Cox Bay they begin to feel the bottom. They hesitate, crowd together and rear into the air, mounting higher and higher as they move toward the beach.

Encountering a rock or reef, waves trip and sprawl headlong, hurling masses of water across the jagged surface in a furious torrent. I see the deadly danger. A body in that water would be tumbled like a rag doll across fangs of stone, savaged and torn. In past times, when the bodies of shipwrecked sailors were found, they often lacked heads and extremities, the appendages having been detached by the surf.

Inside the reefs, the entire surface of the bay is swept by long transverse ridges of water, steep waves moving inexorably toward the beach. As the seabed slopes upward and the water becomes shallower, wave after wave topples forward and closes out. The last hundred metres before the beach is a chaos of leaping waters.

By now I am thoroughly wet, battered by wind and numb with cold. I am also, truth to tell, feeling more than a little anxious. I realize too late that I have strayed out of my depth. I'm unnerved by the tremendous forces raging around me — so impersonal and uncontrolled — and by my growing sense that with some slight care-lessness I could die in this place. This storm, without pausing one iota from its other business, could convert me to a statistic, an item on the evening news.

This is no idle fear. There's a reason the rock around me is bare of vegetation. From time to time the surf surges across this spot and sweeps it clean. Stories abound of surprise waves, much larger than the others, coming from nowhere to pull the unwary from shore. Rogue waves are a real phenomenon. A number of smaller waves coming from different directions, travelling at different speeds, fall into step for a moment, combining forces before going their separate ways. The wave that results is a temporary sum of the smaller waves. If that were to happen now, just there, the sea would climb this rock and eat me alive. I would be washed from my perch, drowning in the foam, lost without hope.

Even so, I can hardly bring myself to retreat. It is wildly beautiful here, a feast for the senses, full of movement, sound and fury. It is a waterfall thirty kilometres wide, stretching from horizon to horizon, blessedly free of any least shred of human artifice. I'm enthralled.

This is the hard edge of the untameable universe, authentic and real, intruding against my familiar day-to-day. It gives me immense satisfaction to experience it. I am persuaded that the sea will be for-ever wild, as wild as the roots of mountains, as wild as the far side of the Moon. This raging sea is beyond the ability of governments and corporations to manipulate or tame. It could be poisoned, I'm sure, filled with garbage, made sterile. Human beings might steal its wealth while it sleeps, perhaps even inhabit it. But we will never tame the sea, never master it, never bring it to heel. Here is one part of the natural world safely beyond our reach.

This comes as a great relief. Here is something timeless, something to depend on. I love the idea, which comes to me now, that in a thou-sand years, ten thousand years, a million years, when time has carried away even the rocks I stand on, there will be storms just like this one blowing in from the deep ocean.

Lennard Island is embattled today, half obscured by rain and blowing spray. The swells break high over its rocks. The reefs are invisible. I see only explosions of surf, an erratic, volcanic violence, a glacial eruption of molten ice. The buildings are dark; there is no sign of life.

Then the beacon blinks, pierces the gloom, cuts like a needle. I wait, nine, ten, eleven seconds, and it blinks again, regular and reliable. Here is the moment of glory, the grand purpose fulfilled. Here it stands fast, the guardian, a bastion marking the very edge of human domains against the dark, curling ocean. *What lies beyond is utterly alien*, it proclaims. *Proceed at your peril.*

> *Sécurité! Sécurité! Sécurité! West coast Vancouver Island north: storm warning upgraded to hurricane force wind warning. West coast Vancouver Island south: gale warning upgraded to storm warning.*
>
> — Environment Canada, *Marine Weather Forecasts*

February

AXHÅML

(Bad Weather Moon)

Winter loses its way in February. There is still much unpleasantness to come and usually at least one major storm sometime early in the month, one last kick at the can. Just when we begin to think that spring is at last on the way, the dawn comes cold and grey and ugly. But winter's power is ebbing, losing steam. Little by little, the west coast of the island slides into a netherworld, not quite winter, not quite spring, where we'll spend the next two months. Sunny breaks become more frequent and powerful. The sun rises farther and farther north along the mountainous horizon to the east. The days are noticeably longer. The birds begin to pair off: eagles, crows, ravens, oystercatchers.

In old times, hope must have revived in late winter. The worst was past and the people, well settled in their spring villages now, watched for signs of an early herring run. Other creatures gathered too, waiting. Everything, it seems, eats the herring or herring eggs, or both. Wheeling flocks of frenzied gulls heralded the great schools, so thick that a fisherman could quickly fill his canoe using a dip net or a herring rake: paddle-shaped with long, sharp spines projecting from both edges of the blade. The rake was swept edgewise, back and forth through the school, impaling fish on the spines. Fishermen also trolled with baited hooks for enormous chrome-silver spring salmon, which hunted the herring where they massed in the bays and coves.

In modern times, too, people relax a little when February arrives. Anyone who went away, stays away — nobody comes back to February if they can help it — but those who have toughed it out through a west coast winter can start to reap the rewards of their fortitude. The busy traffic of summer is still months away. There is time to visit friends or venture into the countryside on small outings. The real fierceness has gone from the weather and there is a euphoria in the blue-sky breaks that not even summer can match: crystalline air and no fog to spoil a promising day; the dramatic juxtaposition of sun and storm.

Meares Island
Lemmens Inlet

CLEAR AND COLD: A Break in the Storm

The Third Street vista affords a glorious view of Tofino harbour. The thickly forested slopes of Meares Island across the water rise steeply from tide line and the snow-capped peaks of Vancouver Island stand bold against the blue sky. The sheltered interior of Clayoquot Sound is rich with the sort of picture-book scenery that's typical of coastal British Columbia. Dark forests loom above narrow saltwater channels. Deep inlets thread their way into the heart of Vancouver Island. It's a heroic landscape, like something out of a Norse saga, but with that special indefinable flavour of Canada's Pacific coast.

Today, unfortunately, that view exists only in my imagination; anything more than a hundred metres above sea level is lost in the clouds. A gale from the southeast is raising whitecaps in the harbour. The little islands are blurred by rain and mist. It's all very discouraging. My friend, the artist Mark Hobson, has offered the use of his floathouse for a couple of nights, starting tomorrow, while he delivers a large canvas to an eager client in Vancouver. The floathouse is anchored at the north end of Lemmens Inlet on Meares Island, in a little cove that rejoices in the name of God's Pocket, a lovely place just over eight kilometres from Tofino as the raven flies, three hours, easy paddling. But just now, in this weather, the idea of paddling anywhere, even just a few kilometres up Lemmens Inlet, seems thoroughly uninviting.

The room is dark, but the reading lamp casts a splash of bright light across the clutter of marine charts and topographical maps on my desk. From time to time, a gust of rain rattles the windowpanes. The spruce outside, battered by yet another southeast gale, thrashes back and forth against the house. Since my visit to Frank Island, the weather has been unrelenting: one storm after another.

The gear is ready. Preparing for the first paddle of the season, especially an early trip in poor weather, is a protracted exercise. This southeast gale will funnel straight up Lemmens Inlet and I expect a cold, soaking wet paddle, not without some hazard. Everything is double or triple wrapped in rubber, vinyl, plastic. God bless the petrochemical industry. The portable VHF radio is freshly charged. My flares and all the other items in the emergency kit are checked and ready for action.

I've been wondering all evening if I should even bother. It's not going to be much fun. The smallest excuse would get me off the hook. The question is this: at what point can I call the whole thing off and not feel like a quitter? It's easy to make decisions at the extremes. If Coast Guard Radio is predicting storm-force winds by morning, I could back down with honour. Nobody in their right mind goes out in such weather. But what if they merely forecast continued gales? Or downgrade it to strong winds? When is one allowed to decide that enough is enough? Where is that fine line that divides discretion, the better part of valour, from a simple reluctance to bite the bullet and get on with it?

It's all moot in the morning. When I peer from my window, the sky is perfectly clear, the sunrise unmarred by a single cloud. The logs along the top of the beach are white with frost; so, too, the tall stalks of dune grass. The air is absolutely calm. Smoke oozes from my neighbour's chimney, drifts down the side of his house, flows gently toward the ocean.

On the morning weather report, bulletins from around British Columbia paint an interesting picture. In the central part of the province, temperatures average around minus 20°C. Locations in the Lower Mainland, including the city of Vancouver, report strong out-flow winds and snow flurries. A blizzard rages in Nanaimo on the east side of Vancouver Island. Travel is not recommended.

In sharp contrast, most stations along the west coast of Vancouver Island are reporting perfect visibility and light winds from the east or northeast. It's one of our classic mid-winter breaks from the weather, blessed relief just in time for my trip.

Every so often during the winter months, a continental Arctic air mass out of Alaska and northern Canada surges southward across mainland British Columbia. The mass funnels through gaps in the Coast Mountains, generating fierce winds as it pours through the narrow mainland valleys and rolls out across the Strait of Georgia. If the mass has enough momentum, it climbs the east slopes of Vancouver Island, spills through the passes and flows down the western slopes.

At such times, a sort of reverse Chinook develops. (Chinook is the warm, drying wind of the Canadian Prairies.) The air mass cools as it climbs the east slopes of Vancouver Island. Water vapour lifted from the Strait of Georgia condenses into rain or snow. By the time the Arctic air reaches the height of land and starts down the west slope, much of the moisture it was carrying has been wrung out. Now here's the thing. Drier air warms more rapidly than moist air as it com-presses. By the time the air mass reaches sea level at Tofino, it is both drier *and* warmer than it was at sea level in Nanaimo. This is the adi-abatic effect.

On the west coast of the island, we notice the wind shifting into the east or northeast. Then it dies away to practically nothing. Skies clear and we're treated to a day or a week or even a couple of weeks of beautiful sunny weather. It's like summer, except that Chesterman Beach never enjoys such clear air in summer. While Tofino basks in warmth and sunshine, Nanaimo suffers bitter cold and blowing snow. Sweet justice. We get to be in the rain shadow for a change.

By the time I finish getting myself organized and onto the water, noon hour has come and gone. The harbour is perfectly calm, except for a few minor ripples where the tide runs more swiftly through the narrower channels. When the light falls just so, the liquid surface takes on the lustre of molten metal. The word "molten" seems entirely apt. This water is so cold — perhaps eight or nine degrees Celsius — that one might reasonably think of it as molten ice. I'm wearing a life jacket over my wetsuit and I'm confident that I could manage fairly well if I were to capsize within reasonable distance of shore. But I'm under no illusions. If I came to grief on some extended crossing and was immersed for any real length of time, hypothermia would soon finish me off.

An early-season trip can be a memorable experience. To begin with, you have the country pretty much to yourself. This morning I seem to be the only kayaker on the water. In summer, so close to Tofino, there would be paddles flashing in every direction. And the view is exceptional. The air is transparent, crystalline in a way that's unique to these mid-winter breaks. Sunshine sparkles from the water. The little islands all around are blanketed with verdant greenery, a wondrous thing in February. Birds are everywhere. A squawking heron lifts himself into the air from the tide line where he's been hunting dinner. I feel very much at one with the natural world around me.

On the other hand, my wetsuit is rapidly becoming an infernal exercise machine, resisting movement every which way. For winter paddling, I wear a suit made for cold-water surfing, thicker neoprene on the body for warmth, thinner layers on the extremities for movement. At least that's the theory. In practice the compromise serves neither purpose particularly well. I end up feeling both cold and encumbered. What should be a gentle, pleasurable outing is rapidly becoming a sentence of hard labour. And it's cold out here, even in the sunshine. My fingertips are numb, despite the insulated rubber gloves.

Perhaps I should have arranged motorized transport to the float-house. I could have borrowed a boat or chartered a water taxi. I considered those options but deliberately set them aside, having done some serious thinking on the subject of transportation after the helicopter flight to Lennard Island in January. To avoid any taint of self-righteousness, I should confess that, as a closet technophile, I enjoyed the ride very much. Too much. Regardless of my opinions on ecological issues, pollution, wasteful consumption of resources and so on, I feel an undeniable attraction to anything shiny, fast and full of gauges. Transportation machines have a fabulous and seductive quality. They are the modern-day equivalent of Scheherazade's flying carpets, bearing their riders to exotic, faraway places in ease and comfort.

Unfortunately, speed, convenience and comfort come at a price. When we travel in motor vehicles we alienate and isolate ourselves from the natural world. Technology is an impediment to intimacy. We arrive as intruders, foreign objects, and the natural world recoils. Sensible creatures run, hide, abandon all their ordinary business to focus anxious eyes and ears on the sudden apparition. Motor vehicles are extensions of the artificial environments human beings have constructed for themselves, little pods of mobile habitat. When we travel in a machine, we never really leave town.

The other objection to mechanized transport is that it can actually restrict mobility. Motor vehicles can't go everywhere, glossy magazine ads notwithstanding. Most are excessively fussy about the conditions under which they will or will not operate. Here on the west coast of Vancouver Island, even a four-wheel drive needs a clear road and, arguably, anything at the end of a clear road is not worth seeing. Boats need a clear channel. Aircraft need clear air and a place to land.

Soon the machine, like a bad-tempered lover, starts to dictate our itinerary, threatening a tantrum whenever we suggest a destination it doesn't care for. Little by little, to save trouble and keep the peace, we fall into the habit of going only where the machine, lord and master, wants to go.

With all that in mind, I've decided to avoid high technology wherever possible from now on. At least I won't be hostage to it. If I want the best experiences, if I want to minimize the gap between myself and the natural world, I'll have to get by with a minimum of gear. The rigorous approach. The price of my pride will be discomfort; I expect to get cold, wet and weary on a regular basis.

I've chosen the easternmost entrance to Lemmens Inlet, a narrow channel between Meares Island and Morpheus Island at the mouth of the inlet. North of Morpheus, the channel opens onto the broader waters of the inlet. From there I'm treated to a full panoramic view of Meares Island. It's a fair size, about fifteen kilometres long and almost as wide, shaped like an irregular horseshoe, toe to the north, the hollow space in the middle of the horseshoe being Lemmens Inlet. The eastern shank of the horseshoe is dominated by the bulk of Mount Colnett, forested and craggy, rising steeply from the water on my right. The equally dramatic shape of Lone Cone Mountain dominates the western shank. The inlet curves gently out of sight, far end still hidden. But I can see the low hills at the northern end of the island and the high mountainous country of Vancouver Island beyond, dominated by the spectacular peaks of Mount Mariner, wearing a fresh blanket of snow this morning. A dramatic landscape, freshly thrilling every time I see it. One would be hard-pressed to find a better view.

Sitting quietly, tucked into the rocky shoreline enjoying the view, I'm reminded, as always, of the exotic flavour of the scenery here. Its oddness. In many ways, this hardly seems a Canadian landscape at all. The abrupt conical peaks rising from salt water, the strange riotous greenery, the irregular skyline — all rather different from anything you might encounter in the Rocky Mountains, the Coast

Mountains, or even the mountains of interior Vancouver Island. If the air weren't so chilly, I could almost imagine myself in Polynesia.

Perhaps that's what early settlers had in mind when they concluded that Lone Cone was an extinct volcano. It does look like something out of *Treasure Island* or *The Swiss Family Robinson*. But Lone Cone's conical shape was not raised by flowing lava. It was sculpted — by flowing ice.

Nearly all of this landscape's defining elements — the long narrow inlets, the steep-sided mountains, the extensive sandy beaches along the outer coast — are souvenirs of the ice ages. On this chilly day, high mountains gleaming with fresh snow, it doesn't take much imagination to picture the glaciers on Mount Mariner growing and growing, spilling down the mountain into the Bedwell River valley, coalescing with the ice flowing from neighbouring peaks, surging toward the open ocean.

The Bedwell River valley and Bedwell Sound line up almost perfectly with Lemmens Inlet. I wonder if the glacier oozing down Bedwell Sound didn't simply climb up and over the low rocky hills at the north end of Meares Island — the way a river flows up and over a sunken reef — then down Lemmens Inlet, across the spot where I'm resting, over Tofino, and out into the ocean.

Back at home I have an aerial photograph of the great Kaskawulsh Glacier, 65 kilometres long, flowing through the Icefield Ranges of the western Yukon. The distant mountains are all but buried in a great swelling field of ice, as much as a thousand metres deep. Only the highest peaks show above the blanket of white. That photo is a time machine. I could be looking at Bedwell Sound, Alberni Inlet or any other inlet along the west coast of Canada or Alaska at the height of the last ice age.

Glacial history is necessarily obscure. Each fresh advance of the ice sheet ploughs up evidence left by smaller previous advances. (Not a bad metaphor for life, when you think about it: who cares if you muffed Physics 12 or made a fool of yourself on your first date, so long as you didn't retreat for good afterwards?)

Geologists believe that glaciers advanced many times during the most recent geological era, the Pleistocene. Pleistocene glaciation falls into four major episodes — the Nebraskan, the Kansan, the Illinoisan and the Wisconsin — irregularly spaced over the last million years. Each major episode represents a cluster of several distinct glaciations and each glaciation comprises several cycles of advance and retreat called stades and interstades. The most recent Pleistocene episode, the Wisconsin, occurred between 128,000 and 7,000 years ago.

There is evidence along the west coast of Vancouver Island for at least three different advances of glacial ice, all from the Wisconsin Episode. The most recent of these, some 11,000 years ago — the Sumas Stade, Fraser Glaciation, Wisconsin Episode — was a relatively minor advance. Glaciers probably never escaped the mountain valleys. But during the Vashon Stade of the Fraser Glaciation, between 13,000 and 21,000 years ago, the edge of the ice sheet advanced to a point about six kilometres west of the present coastline. The place where I'm resting was buried beneath one hundred and fifty metres of grinding, creaking glacial ice.

That was not the all-time maximum. At the height of an earlier advance — during the Salmon Springs Glaciation of the Wisconsin Episode, 37,000 years ago — glaciers extended 12 kilometres over the ocean and buried the mouth of Lemmens Inlet to a depth of 460 metres. In the mountains of central Vancouver Island, the ice was one and a half kilometres deep.

So imagine yourself on a calm summer day, floating in your kayak fourteen or fifteen kilometres off Tofino. One kilometre away an ice cliff stretches north and south, seemingly endless. You dare not paddle closer. The air resounds with the crackle and roar of icebergs calving from the glacier into the Pacific. (We don't have to imagine that; it's familiar to anyone who has visited Glacier Bay, Alaska, or seen Lemaire Channel and Bransfield Strait in Antarctica.) Beyond the ice cliff, the looming slopes of the glacier, brilliant in the sun. The nearest mountains are all but buried in ice. The peaks of Colnett and Lone Cone show only as little rocky nubbins, called nunataks, standing clear of the ice sheet. Still the frozen surface slopes upward toward the heart of Vancouver Island. Back in the Bedwell River valley, beside Mount Mariner, the glacier is over a kilometre thick.

The whole great mass, plastic under its own enormous weight, oozes toward the ocean — at a glacial pace, to employ a useful cliché. As it moves, it plucks frost-loosened rock from buried valley floors and slopes. More rock falls from whatever mountain faces remain exposed above the glacier. All that rock is carried along in the ice like so much grit on a gigantic piece of sandpaper. The rock-laden glacier grinds and shapes every surface it touches. Like a great conveyor belt, the flowing ice carries huge volumes of excavated rock, sand and mud toward the coast. There, at the melting edge of the glacier, termed the toe, all that debris is finally freed from the ice into a growing mound of guck: the terminal moraine.

Hence the odd profile of these glacial inlets, hollowed out by rasping tongues of ice. Typically the deepest parts of the inlet are far inland, where the grinding and excavating were most intense and prolonged. The inlets of Clayoquot Sound are more than 160 metres deep in places. Alberni Inlet — at 56 kilometres, the longest inlet on the west coast of Vancouver Island — boasts depths of more than 350 metres. But glacial inlets are often surprisingly shallow at their mouths, over the terminal moraines. The bar across the entrance to Tofino harbour is less than six metres deep at low tide.

As the climate warmed at the end of each glacial episode, ice began to melt at an accelerated rate. As the mass of ice dwindled, glaciers ceased to flow. It is commonly said that glaciers retreat, but the expression is not apt. A glacier in decline simply ceases to move, paralyzed by changing circumstances, and melts away where it lies, like the Wicked Witch of the West. The mass of sand, gravel, rock and mud imprisoned in the dying, immobile glacier is released as the ice melts.

Such material has a characteristic appearance quite different from material transported by flowing water. Water-borne rock is rounded; all the corners get knocked off as the stones tumble downstream. Water-borne material is also well sorted according to size. The biggest rocks are dropped in one pile, gravel in another, sand in yet another, all depending on the speed of the current at various places in the stream. But material that falls out of a dying glacier, called glacial till, is irregularly shaped and absolutely unsorted. Great boulders are mixed willy-nilly with gravel, sand and minute particles of silt, just as they came out of the rotting ice.

All of the low-level country between Tofino and Ucluelet is one enormous deposit of glacial till, part of what geologists call the Estevan Coastal Plain. Post-glacial streams running through this deposit have started the long job of sorting the material. They cut meandering valleys and scour away the mud and silt, leaving larger rocks behind.

On the seaward side of the coastal plain, the same process is under-way. The ocean bites into the coastal plain and the surf carries off the finest materials, leaving the larger stuff behind. The lovely beaches for which the area is renowned are made of sand washed free from glacial till. The process can best be seen in action where the ocean is, even now, eroding the bluffs behind Wreck Beach on Florencia Bay.

So the landscape of western Vancouver Island owes much of its present appearance and flavour to ice-age events. The long, deep, rel-atively straight inlets with their hanging valleys; the bowl-shaped cirques, eroded ridges and Matterhorn spires of the mountains; the relatively level floors and steep walls of the valleys (U-shaped rather than V-shaped) — these are all classic glacial features.

Even the little rocky islet beside my kayak shows the characteristic grooving etched by glacial ice. Somehow that brings the whole busi-ness alive for me. I've seen exactly the same sort of grooves on bedrock newly exposed under shrinking glaciers in the mountains of south-central British Columbia. The polish is gone from this rock, worn away by millennia of saltwater weathering, but the glacial shaping is still clear and exquisitely evocative of past ages.

Out of an abundance of glacial silt and through the magic of estuarine chemistry, Clayoquot Sound is gifted with large areas of mudflat. Wherever significant amounts of sediment-laden fresh water mix with salty, nutrient-rich seawater, chemical and physical interactions cause minute particles to precipitate. Mudflats are among the world's richest natural ecosystems, outstanding habitat for wildlife, accommodating everything from sea ducks and gray whales to sculpins and ghost shrimp.

One of Clayoquot Sound's largest mudflats, the Arakun, lies just ahead. At high tide the whole thing would be covered with water and I could paddle straight across. But just now, exposed by the ebb, the mud extends out a couple of kilometres from the eastern shore of Lemmens Inlet. Only a narrow channel remains open along the western shore, far off to my left. It's dawning on me that, having chosen this easternmost entrance, I'm going to have to double back the whole width of the inlet. What was I thinking? Ah well, it's all part of the experience.

There's a breeze now, just a gentle drift of air down the inlet, but cool. Time to get moving. As I paddle toward the channel, the view opens up northward. I can see the far end of the inlet and the mountain peaks along either side of Bedwell Sound. I start to encounter large flocks of waterfowl riding on the water: Canada geese, mallards, buffleheads, goldeneye. A small flock of western grebes fishes the channel. These grebes are fair-sized birds — larger than ducks, smaller than geese — and elegantly attired, dark above and gleaming white below, dressed for dinner with white tie and tails. There are little groups of mergansers, both common and red-breasted, similarly occupied. Several cormorants fly by, species unknown, looking ungainly in the air. A line of airborne surf scoters whistles past, medium-sized diving ducks, compact and black.

I can see and hear many more birds out on the mudflats. There is even a group of killdeer somewhere out there. I can hear them — old friends that I associate with the dry interior of British Columbia. They do nest sometimes at Tofino airport, but I expect these are tourists. I wonder where they will find themselves come spring. Bunch grass and Ponderosa pine country, perhaps, or some farmer's pasture. Farther away, on the other side of the inlet, I can just make out a couple of large white specks and I wonder if they might be trumpeter swans, which occasionally turn up during the winter. There are gulls everywhere: Thayers, glaucous-winged, mew gulls.

Oddly enough, deep winter is the best time to see wild birds in Clayoquot Sound. Heavy snow forces most of the passerines out of high-elevation forests, concentrating them along the seashore. Waterfowl flock to the sound between November and February to escape bitter weather in the rest of the country. British Columbia boasts Canada's best habitat for wintering birds and the inlets of Vancouver Island play an important role in providing for feathered transients.

The mudflats seem to go on forever. I follow the shoreline past Arakun Islets and back toward the eastern side of Lemmens Inlet. The ebb has slacked right off; the tide is turning. It's perfect timing. I'll have the current with me all the rest of the way. At low tide the flats are high and dry, relatively speaking. The ground looks almost solid. I'm tempted to land and get out for a bit of a stretch. But I know from experience that I might find myself knee-deep in mud.

All things pass. Eventually the Arakun falls astern. Farther up, the inlet is more typical of glacial valleys, rocky bluffs rising from deep water. This is extremely rugged country, a fact belied at some little distance by the thick covering of forest. Up close, the mountains reveal themselves as a series of nearly vertical faces, cliff upon cliff. What appears to be a continuous fabric of vegetation is really a patchwork, little clumps of trees anchored in small accumulations of soil among the cracks, terraces and ledges, all clinging tenuously to the craggy face of the mountain. It's amazing that a forest can take root, grow and prosper on such meagre substrate and across such steep slopes.

Incidentally, the steep walls of Lemmens Inlet are made of very different material from the pinnacle that took a chunk out of my hand at Frank Island last month. The bedrock there was mostly sedimentary: dark fine-grained argillite or mudstone alternating with layers of pale chert. The bedrock in Lemmens Inlet is mostly granite-like. Even without a magnifying lens, I can see the crystalline minerals in an attractive contrasting pattern of light (quartz and feldspar) and dark (hornblende).

The bedrock along the inlet is also very much older than the bedrock at Frank Island. In fact, this is some of the oldest rock on Vancouver Island. A sample collected at Grice Bay was approximately 260 million years old. Mostly igneous or metamorphic, it belongs to what geologists call the Westcoast Crystalline Complex, originally formed of molten matter deep in the Earth's crust. The sedimentary rocks of Frank Island, practically next door, are less than half as old, 120 million years or less.

The terms "plate tectonics" and "continental drift" have come into common use, but the theories they represent revolutionized geological thinking during the 20th century. It was Alfred Wegener who proposed in 1912 that present-day continents represent the dispersed fragments of a single ancient supercontinent, Pangaea. In the 1930s, Arthur Holmes proposed that thermal convection in the hot, plastic material of Earth's mantle might be the mechanism by which continents are broken up and forced apart. But the whole theory received little attention until the 1960s, when study of geomagnetic anomalies suggested that new bedrock was being added to the crust along mid-ocean ridges. Howard Hess and R. Deitz called the phenomenon "sea-floor spreading."

The Earth's crust is not a solid uniform covering like the shell of an egg. It is, rather, a fractured mosaic of different-sized chunks, called plates. These plates float on Earth's molten mantle and are in constant motion, grinding and jostling like cakes of ice on a wintertime river. Sometimes the plates draw away from each other, opening a rift to be filled with molten material from below. Sometimes they collide, so that one plate is thrust beneath another, a process called subduction. The edge of the subducting plate is forced down into the mantle to be recycled.

It's a messy process. Subducting plates do not go cleanly into the fire. Bits and pieces, mostly surface material, end up plastered along the edge of the overriding plate, like mud sticks to the blade of a bulldozer. Any large mass projecting from the subducting plate, an island or a mountain, might also break free and be left behind. If the projecting mass is large enough, the whole subducting plate might jam firmly against the overriding plate, so that the two become one mass. All these various-sized fragments of captured exotic crust are called terranes.

How does all this apply to Clayoquot Sound, Frank Island and Lemmens Inlet? At present, the North American Plate is drifting westward. The leading edge, the business end of that plate, lies just 30 or 40 nautical miles (55 to 75 kilometres) west of Tofino. There it meets and overrides the two relatively small eastward-drifting plates that comprise the nearest piece of deep ocean floor: the Juan de Fuca and Explorer plates. Farther out, the much larger Pacific Plate is drifting northeastward toward Alaska and the Yukon.

The process has been underway for a long time. The Juan de Fuca and Explorer plates were once much larger. And there were other plates, now completely vanished, whose only vestige is found among the jumble of material accreted onto the western edge of the North American Plate — the rocks and mountains of western British Columbia. In fact, better than half of this westernmost province of Canada is made of exotic terranes jammed onto the main continental mass during its westward drift.

According to current theory, the larger part of Vancouver Island belongs to a terrane called Wrangellia, which also includes the Queen

Charlotte Islands, farther north along the coast of British Columbia, and the Wrangell Mountains of southwestern Alaska. Wrangellia took form about 380 million years ago as an arc of tropical volcanic islands far out in the ancestral Pacific Ocean. In the course of 100 million years or so, these islands eroded to broad submarine shelves, home to a prosperous growth of marine animals. Countless generations of accumulating shells and coral eventually formed thick layers of limestone. Then, about 230 million years ago, the island arc was split by fractures. Thick sticky lava flowed onto the surface, layer upon layer, accumulating to a depth of many kilometres. After that came more marine growth and still more volcanism, including the formation of great granite batholiths that would become the peaks of Vancouver Island.

Meanwhile, the North American Plate, then in the process of separating from Europe, was advancing westward toward Wrangellia. About 100 million years ago the two masses collided, with consequences felt as far away as the foothills of the Rocky Mountains. Both the continent and the incoming terrane were severely bent out of shape, compressed, buckled, uplifted and eroded. The heat and pressure were sufficient to partially melt and re-form rock, changing limestone to marble and granite to gneiss.

But the story doesn't end there. Around 40 million years ago, two much smaller terranes, the Pacific Rim Terrane and the Crescent Terrane, jammed themselves under the outer edge of Wrangellia. The Pacific Rim Terrane, smaller of the two, mostly sedimentary — a fragment of the North American Plate that had somehow come adrift — now comprises the extreme outer edge of southwestern Vancouver Island. The mostly volcanic Crescent Terrane, formerly a largish piece of seafloor, is now represented by the extreme south tip of Vancouver Island and the Olympic Mountains of Washington state across the strait of Juan de Fuca.

The bedrock of Frank Island belongs to the Pacific Rim Terrane. The bedrock along the shore of Lemmens Inlet belongs to Wrangellia. In fact, the bedrock around Lemmens Inlet belongs to the deepest, most ancient layers of Wrangellia, bent upward by the force of the collision with the Pacific Rim Terrane and later exposed by erosion.

The actual fault line where the Pacific Rim Terrane meets Wrangellia must lie somewhere between Meares Island and Frank Island, mostly buried under glacial debris. I've spent some time searching for a glimpse of it on the surface, without success. It might be worth seeing — a physical border marking what was, for 60 million years or so, the ancient edge of North America.

One of the problems with geology, with natural history in general, is that it's far too easy to fall into a habit of seeing the history of the world in terms of grand abstractions. Mountains rise, terranes come crashing ashore, whole continents sail across the surface of the globe, species of plants and animals come and go. We come adrift from reality. The antidote to that kind of fuzzy, big-picture perception is to think about the lives of individual actors. That's where the real charm of natural history lies, anyway. Rocks, like people, have curricula vitae, life histories, stories to tell.

Back home on my desk is a piece of argillite from an outcrop near Frank Island, one little bit of the Pacific Rim Terrane until I pried it free and carried it away. The little particles of sand and silt in that rock settled out of the Pacific Ocean 120 million years ago, not far to the south, somewhere along the continental slope in the neighbourhood of present-day Washington state. Even then, those little grains represented the end product of an unimaginably lengthy process. They are the dust and bones of extinct mountains first reared to the sky by time and a restless Earth, then erased with the same casual ease. Over eons, the pulverized remains of those mountains, the fine particles of sand and silt, rained down out of the ocean together with the remains of marine plants and animals, building layers of seafloor mud that alternated with layers of silica gel. Tremendous depths of sediment accumulated. Eventually the deepest layers petrified under the pressure, becoming part of the bedrock — the ooze becoming argillite, and the silica, chert.

For whatever reason, perhaps because of a shearing motion from adjacent plates, the newly formed piece of the continent, including the rock on my desk, broke loose, drifted northward, then jammed solid under Vancouver Island about 40 million years ago. What followed seems almost trivial by comparison: dramatic reshaping by repeated glaciation over the last four million years, then a few millennia of erosion by wind and wave to excavate soft argillite from between the layers of hard chert.

The mind boggles at the span of time it took for grains of rock to erode from the primordial continent; to be carried metre by metre down some forgotten river into the primeval Pacific; to settle on the sea bottom and lie there long enough to be covered by thousands of metres of further sediment; to petrify into new rock and, in this second life as sedimentary rock, to break away and be carried northward on the back of a crustal plate; to be plastered onto the edge of North America in a monumental fender-bender; to be thrust up into the air, carved by glaciers and, almost as an afterthought, eroded into jaggedness by the surf.

We get far too casual dealing with such large numbers, an eon here, an eon there. It's all much too facile. We need some perspective, a new anchor. We need to remember how grindingly slow the process really is. Think about Frank Island again. Think about this last and least part of the story. How long does it take the ocean, splash by splash, grain by grain, to wear down those layers of argillite? How many fractions of a millimetre per century? Would we see any change in one human life?

The rest of the trip is uneventful. The weather holds beautifully, but the cold is getting inside me. I'm relieved to fetch the entrance to God's Pocket at last and escape the wind. It could be worse. Lemmens Inlet, blessed with a relatively complicated and interesting shoreline, is not truly representative of glacial valleys elsewhere in Clayoquot Sound. Here there are lots of little coves to explore; I could play hide and seek with the wind. Bedwell Sound just over the hill is twice as long and offers practically no shelter at all; glaciers have ground it straight and plain, like an enormous ditch. Paddling can be both monotonous and terrifying — dangerously exposed — especially in a winter gale or on a summer afternoon when the onshore wind starts to blow. I don't suppose it would have been all that dangerous today, but it would have been a long, hard paddle against the wind.

Such a pleasure to arrive at the little floathouse, delightful place. I've been here many times, in all sorts of different weathers, and the first glimpse through the narrow entrance to the cove never fails to cheer me. In summer, it's a hanging garden of potted plants; in winter, a cozy haven, especially if smoke is curling from the chimney. Part of my pleasure stems from the incongruity of it all: in this wild place, a little wooden cottage floating on the ocean, like something out of a fairy tale.

I'm conscious of the irony in my own reaction to the scene, my enthusiasm. I've come out in search of wilderness experience, but I'm only too glad to abandon it all for the promise of a cozy shelter at journey's end. On the other hand, shelter is rapidly becoming a genuine issue. The days are still short in February. I've taken my sweet time on the journey, there have been delays and detours, now it's getting late. What little warmth the sun has generated is already fading. My hands are clumsy and blue with the cold. I'm chilled to the bone. Getting out of a tight-fitting kayak and onto a floating dock is a tricky business at the best of times. Moving awkwardly, in slow motion, I manage to extricate myself without mishap, lucky not to end up in the water.

Getting out of the wetsuit is also a big production with fingers that won't do as they're told. At length I'm free, a blazing wood stove is pumping heat into the floathouse and a kettle is starting to sing. Whenever I get in from a long paddle in summertime I like to go for a swim to freshen up, but the idea doesn't appeal today. I'll make do with a hurried sponge bath on the dock.

I'm just tipping what's left of my warm water into the bay when I notice something odd. Just under the surface, next to the floating dock, a single hemlock needle drifts slowly, steadily, toward the mouth of the cove. Twenty centimetres down, a fragment of kelp moves steadily in the opposite direction, toward shore. This is the sheltered side of the floathouse. The water is perfectly calm, undisturbed by the slightest breath of air. When I crouch down and sweep my hand back and forth, the layer of water at the surface takes on an oily, shimmering, heat-haze quality.

Intrigued, I fetch a cup from the sink and fill it carefully, just dipping the rim below the surface. After a moment's hesitation, I sip. The water in the cup has only the faintest brackish taste. Hardly salty at all. I could probably drink it. Now that I'm paying attention, I can hear the murmur of many little streams percolating down, through and over the gravel beaches around the cove. Runoff from the heavy rain of the past few days is still pouring from the mountains and flowing *onto*, not into, the ocean. All that fresh water is spreading across the calm surface of the bay in an unbroken sheet about 15 centimetres deep.

I've read about this phenomenon, but how odd and vivid it seems in real life. There are really two bodies of water here, stacked one on top of the another: a freshwater lake over a saltwater inlet. I know that salt water is denser than fresh water. That's why swimmers find it easier to float in salt water. That's why salmon entering fresh water to spawn must gulp air to increase their buoyancy. That's why overloaded boats, barely afloat in the ocean, sometimes sink when they enter a freshwater river or lake. That's why fresh water floats on salt water.

In confined waterways like God's Pocket and Lemmens Inlet, the layer of fresh water flowing toward the open ocean will capture a certain amount of salt water from the layer below and carry it away. To replace what is lost, more salt water flows into the inlet from the ocean. That's why the two bodies of water, salt and fresh, are moving in opposite directions.

The whole phenomenon, the important influence of fresh water in these sheltered waterways, is a tidy reminder of just how much the oceanography of the inside passages differs from that of the open coast. Surf, so important on the open coast, is hardly a factor here. Swells never penetrate this far and relatively short reaches hamper the development of big wind waves — though fierce winds funnelling through the mountains can develop some seriously rough water, especially in long straight inlets.

On the other hand, currents are much stronger in the inlets than they are along the outer coast. The ebb and flow of the tide can generate considerable turbulence in narrow passages: rapids, back eddies, whirlpools, upwellings, even saltwater waterfalls. Despite this apparently vigorous flow, the actual volume moving in and out of a deep inlet through the narrow passages can be surprisingly small. The great depths are often stagnant and poorly oxygenated compared to the open ocean.

These sheltered inside waters also have a different aesthetic flavour. Quieter than the outside coast. Sometimes warm and sunny. Sometimes dark and sombre, especially in winter. But if they are more sombre than the outside coast, they are also majestic and spacious. They have something of the mystical about them, rain or shine. The surf and sunshine of the outside beaches makes me want to laugh, run, shout for joy. But dark, quiet valleys on the inside always have me speaking in whispers, as if a bold word might be out of place, an intrusion, like a shout in church. As if something sleeps here it were best not to waken. For fear? Or for fear that, if I'm not careful, something might vanish — as when a flock of swans, startled, takes to the air and is gone, leaving me behind, earthbound and diminished.

I heat more water for instant soup and take the steaming mug outside to enjoy the last bit of afternoon sunshine. The low angle of the sun, the soft light of late winter, has the greenery around the cove almost glowing. Three goldeneye paddle by, unconcerned, while two mergansers work the deep water along the rocky shore. It's profoundly peaceful here, wonderfully quiet. That's something you don't get on the exposed coast, where surf generates a constant, pervasive roar. I hear small birds foraging through the shrubs along the shore and the little splashings made by the mergansers diving for their dinner.

I realize — and this comes as a pleasant surprise — that I'm feeling extraordinarily satisfied, happy, content. There is a wonderful atmosphere of freedom here: freedom from interruption, freedom from unwelcome distractions, freedom from intrusion. It's like coming home. I subscribe to the school of thought that says human beings have a powerful instinctive need for wild countryside, a yearning. I believe also that we suffer a kind of malnutrition or maladjustment when deprived of it.

Perhaps it's simple homesickness. After all, we evolved as nomads, hunters and gatherers. It is only very, very recently that we've abandoned that way of life — in a sense, we're all new in town. Perhaps the warmth I'm feeling just now for this little piece of countryside is simply a hunter-gatherer's perception of promising territory. The ancient forests of Clayoquot Sound are the ultimate in promising territory. Nothing is missing. Nothing has been used up. This cup is still full to the brim.

I'm nonetheless astonished at the power of this humble destination to move me. What a difference a few kilometres of paddling can make. Tofino seems worlds away. The forest, this cove, the evening sky are much as they have always been, true wilderness.

I watch the last glow off the mountaintops before I'm forced to retreat indoors. It's still winter, after all; real spring is months away. Even before the sun is quite gone, the air grows chilly. It's going to be frosty tonight but the stars will be lovely and clear, a rare treat on the coast in winter. Should I stay and watch a little longer? Inside, the wood stove is warm and supper waits. When I step through the door, the hiss of gas in the lamps drowns the faint sighing of wind through the growing darkness outside.

March

AIYAKÅMIL

(Herring Spawn Moon)

March bridges the gap between late winter and early spring. The weather continues chilly and snow is not out of the question. After a storm, the breaking clouds still reveal a fresh frosting of white on the mountains, even if we've had only rain at sea level. On the other hand, the days grow noticeably longer as the sun moves toward the equinox. The first gray whales arrive. Pacific treefrogs commence their lusty chorus. Robins appear and varied thrushes sing their odd two-tone harmonic from the depths of the forest. Skunk cabbage spathes, brilliant yellow, emerge wherever the ground is especially moist. And the first salmonberry blossoms — coming before the leaves — show vivid pink in the undergrowth, looking oddly artificial on still-naked twigs, as if some humorist had decided to decorate the winter woods with imitation flowers. And following the blossoms, our first hummingbirds.

These must have been busy, happy days in the old times. After the lean months of late winter, the supplies of autumn-dried salmon long gone, another season of plenty was at hand. Fishermen continued to take the herring with rakes and nets while the whole community watched for signs of spawning. At the first appearance of milt in the water, weighted hemlock boughs were suspended from floating poles anchored in the spawning grounds. It was a tricky business. False spawning often precedes the main event. If the boughs were placed too early, they would catch only a scanty layer of eggs from the false spawning and the whole setup would have to be replaced, entailing much extra work. With luck, when the water cleared again, the boughs would be covered with thick masses of eggs. Then the whole length of beach in front of the village would be taken up with racks of spawn drying in the sun and wind. Later, when the egg masses were perfectly crisp and dry, they could be packed in baskets or wooden boxes for storage.

In modern times, too, the pace picks up in March and the season of plenty seems suddenly closer at hand. School break brings the first rush of visitors, a rehearsal for a busy summer to come. Entrepreneurs in all the ramifying branches of the area's burgeoning tourism industry put the finishing touches on their businesses and await the arrival of customers. But it's still early days and the pace is relaxed. Whale festival — timed to catch spring break rather than any biological event — is soon over, though whales will continue to migrate past for another month or more. Visitor traffic eases off, leaving local people to enjoy the increasing sunshine and the first stirrings of spring.

Toquart Bay
Barkley
Sound

NATURE'S PLENTY: Herring Time

Wynken, Blynken, and Nod one night
sailed off in a wooden shoe.
Sailed on a river of crystal light,
into a sea of dew.
"Where are you going, and what do you wish?"
the old Moon asked of the three.
"We have come to fish for the herring fish
that live in this beautiful sea;
Nets of silver and gold have we,"
Said Wynken, Blynken and Nod.

— Eugene Field

Early March. Barkley Sound. Deep potholes overflow with muddy water. Rain beats a metallic tattoo across the roof of my truck. Windshield wipers slap a frantic counter-rhythm — back and forth, back and forth, back and forth — barely able to keep up with the downpour. It's a miserable day. But I'm thinking that this deluge from the sky may not be altogether a bad thing. The road to Toquart Bay on Barkley Sound delivers a bone-jarring ride, even with the tires down to half-pressure. But, thanks to all that water, the truck may be hydroplaning over the worst of the gravel washboard. It's an ill wind that blows no good and I'm grateful for any relief I can get. I weave from shoulder to shoulder, proceeding cautiously, trying to pick my way down the path of least resistance.

I am grateful to the rain for another quite different reason: it hides the scenery. The countryside I'm passing through has been devastated — the abandoned Brynnor Mine site, blown-out creeks, whole mountains stripped of trees — an echo of post-industrial landscapes everywhere. What do they call it back east? The rust belt, greening up a little, finally, under the soft influence of time. But here the wounds are still raw and grievous.

Everywhere it's the same deal. Corporations wangle control of the countryside and cream off the resources. After a time, the resources run out and so do the corporations, leaving nature to recover as best it can.

I find myself thinking of similar landscapes just to the north, still mostly intact. I'm only too aware that exactly this sort of dismemberment was planned for Clayoquot Sound, would have occurred long ago, were it not for the environmental protests of the 1980s and 1990s. Nor have I forgotten that there are still corporations, politicians and individuals entirely willing — eager, in fact — to visit the same sort of destruction-for-profit on Clayoquot Sound and who are, even now, actively planning, scheming, manoeuvring toward that goal. The new barbarians.

All in all, I'm in no very happy mood as I roll down from the height of land, headed south, past Maggie Lake and the Maggie River, back to salt water. Barkley Sound is a great southwest-facing embayment of the ocean, dotted with clusters of little islands and hemmed in on three sides by steep rugged country: mountains curtained in mist and rain, ridge upon ridge, fading toward invisibility. There are clearcuts everywhere.

Surprisingly, even thus diminished, the scene retains something of its original majesty, a great sweep and stillness, wild and wet. It calls to mind much of what I felt when I first moved to the outer coast: the attraction of far-off horizons and distant valleys, enormously quiet, the promise, somehow, of absolute freedom, of escape from ordinary cares. I find myself unexpectedly moved — much is gone, but much remains.

At the boat launch, I leave the truck and walk to the water's edge, where little wavelets are lapping the sand. This is winter water, cold and crystal clear. The algal blooms that turn the bay to soup during the summer are still months away. The breeze has a chill bite to it and that lovely saltwater tang. The rain has eased off. Toquart Bay is peaceful and quiet, just the slightest ripple on the water's surface.

I don't even bother to take the kayak off the rack. It's a chilly day for a paddle — fresh snow in the mountains according to the weather report — and besides, there doesn't seem to be much point to it. The schools of herring are out there with the spring salmon, the seals and the sea lions. Leastways, I hope so. But they could be anywhere in the sound, and they could be deep — odds are that I'd never see them. I prefer to stay here on the beach, quiet and content, listening to the waves lap on the sand, watching wreaths of cloud drift among the forested peaks. It is enough to imagine the herring, somewhere out there beneath the dark surface of the water, also biding their time, waiting patiently for the turn of the seasons.

The view was rather different — except for the weather, which was every bit as foul — a few years ago, another day in early March, when I sat on this very spot watching the Barkley Sound seine herring fishery unfold. On that day, Toquart Bay was a scene of frantic activity. Horns sounded, engines revved, men shouted, all that noise echoing loudly across the water. Huge boats moved ponderously among the islands. Seiners are nothing like the size of the sea-going trawlers and factory ships that work the open ocean off the west coast of Vancouver Island. But they certainly *look* big — muscular, predatory, robust, metallic, bristling with technology — particularly when crammed together among the little islands of Barkley Sound. They don't so much float on the water as shoulder it aside. Sometimes they hardly seem like boats at all. They are fishing machines, ocean-going combine-harvesters.

The surreal flavour of Barkley Sound that day was farther enhanced by the dead level surface of the water. The heavy rain had beaten the waves flat and the seiners moved across Toquart Bay like agricultural machinery working a field, or game pieces on an enormous board. The air throbbed with the sound of big marine diesels, deep voices, sea-going locomotives. Below the surface, the water must have resonated to that sound, the voice of doom. Smaller boats hastened back and forth like messengers on a battle field. It's odd how the extraction of natural resources comes to be expressed in terms of a battle, a military campaign, with the natural world as an adversary to be conquered, subdued, plundered — the harvest as tribute to be extracted. On that day, the lines were drawn at the water's surface: human beings versus herring. I had my radio along to eavesdrop on communications across the water. The roll call of vessels had the sound of rough poetry: *Qualicum Producer, Harbour Provider, Ocean Wanderer, Silver Bounty, Western Scout, Sarah Margaret, Silver Viking, Northern Cloud, VanIsle, Nordic Queen, Alaska Queen, Bernice C, Viking Spirit.*

Wherever I looked, there were sets in progress: nets lifted high on booms or winched slowly, slowly onto the great stern drums of the seiners. Swirling clouds of gulls attended the fishermen; sea lions announced themselves with a chorus of grunting, belching roars. Through binoculars I could see the catch being vacuumed — if that is the proper word for a suction hose that looked to be half a metre or more in diameter — out of the nets into waiting fish packers to be salted and refrigerated for the long trip to a packing plant. The ultimate product of the roe-herring fishery is kazunoko, a caviar-like delicacy made from the mature egg skein of the female herring. It is much in demand by Japanese consumers, who will pay $120–$150 per kilogram. Male fish and the remains of female fish are processed to oil and meal. Because the kazunoko has such high value, fishermen can make substantial profit from a relatively modest catch. But it is a demanding fishery. Herring must be taken at precisely the right time: after the eggs have matured, but before the fish begin to spawn. Openings are very brief. Competition for the available quota is intense. Problems with gear or bad weather can spell disaster. And once caught, the herring must be carefully handled to preserve the value of the roe.

To begin the set, a smaller boat — the skiff — carries the leading edge of the net away from the seiner itself. The skiff travels in a huge loop around the school of herring, returning eventually to the larger vessel. The extended net hangs vertically in the water, a curtain encircling the fish. The weighted line through the lower edge of the net, the lead line, is tightened. The bottom part of the net gathers together like a drawstring purse, forming a great rounded sac with the school of herring trapped inside. Then the floating cork line along the top of the net is tightened and the whole outfit gradually winched aboard the seiner. Seawater escapes through the mesh but the herring remain trapped inside. If the catch is sizeable, the fish must be vacuumed or scooped from the net while it is still in the water.

The use of purse seines has always been controversial. When they were first introduced in 1910, they couldn't even be licensed. Other fishermen complained that the seine fishery was hard on herring stocks because it did not distinguish adult from sub-adult fish. The smaller fish — the next generation of spawners, too small to be processed — were destroyed as bycatch. But the demand for herring was growing leap by bound in those days, the industry wanted to catch more fish and, eventually, inevitably, misgivings were talked away or ignored. The first permits were issued in 1913 and seiners have dominated the British Columbia herring fishery ever since.

I watched into the dark of evening. All over the bay, clusters of boats worked their sets, floodlights brilliant in the rain and gathering darkness, fountains and cascades of seawater gushing from the decks and railings as fishermen labouring in the artificial glare continued to vacuum their catch aboard.

It is no easy thing to ride herd on such an intense, high-stakes fishery. Part of the problem is that herring are obligate schoolers. Instinct compels them, absolutely, to gather in groups — schooling is their primary response to stress or danger. The modern seine fishery is terribly effective at finding and catching those schools. With most other kinds of fish, stocks become increasingly scattered as they decline. The catch-per-unit effort falls, as fisheries managers would say: it becomes harder and harder to catch anything. Not so, herring. The remnants of a declining herring stock conveniently gather themselves into dense groups on smaller and smaller fishing grounds. The catch-per-unit effort can actually rise as a stock fails — the fish become easier to catch as the population declines and the harvest actually peaks just before the fishery collapses. Seiners continue to take their allotted catch until the last school is scooped from the water. So the success of the seine herring fishery is no real measure of the health of a stock.

As it happened, the Barkley Sound gillnet herring fishery, ten days later that same year, proved to be a disaster.

The gillnet fishery always follows the seine fishery because it is intended to take herring later in the cycle, as they move out of deep water into the shallows just before spawning. The female fish are that much riper and roe is of better quality, more valuable. On this occasion, the fleet assembled as usual, the nets went into the water and were ready to go, but the herring evidently decided to give the party a miss. Instead of frantic activity I saw idle hands. The radio chatter had an air of mystification. *What's going on? Where are the fish? Is it any better where you are?*

The vessels used in the gillnet fishery are open skiffs, much smaller than the seiners, and the nets work on a different principle. A gillnet hangs vertical in the water, invisible, drifting with the current. When a fish of the proper size swims into the net — the size of the mesh is critical — it jams tight. If it tries to back out, the fine line of the mesh snags beneath its gill covers, preventing escape.

When the time comes to take in the catch, the nets are fed over the skiffs, crossways, from gunwale to gunwale. The mesh comes from the sea spangled with bright silver fish, trout size, hanging like sequins from the web. Motorized power rollers help to feed the mesh smoothly across the skiff while an elongated spinning paddle, the beater bar, beats the net from below, helping to shake the herring free. Fishermen along both edges of the net also shake and pluck and clean the net as it slides across the skiff and back down into the water to catch more fish. The fish bounce like popcorn coming out of the net, making their one last desperate squirming attempt to escape.

A fisherman waved from one of the skiffs: "Great day for a paddle!"

I nodded and waved back: "For fishing, too?"

He shook his head, wryly. "No, not a great day for fishing."

It must have rankled a little that the seiners had taken their full quota ten days earlier. But that sort of thing happens not infrequently. In a limited resource fishery with a huge oversupply of fish-catching capacity, somebody is bound to get the dirty end of the stick. On my way back to Toquart Bay, I came across a couple of men from Macoah village pulling up hemlock boughs that they had anchored in the water, First Nations fashion, to collect herring eggs. The whole bough should have been coated in a thick mass of roe. I saw only greenery. "Any luck?" I asked.

"Nothing," said one. "I don't think there's anything left to spawn. They took it all."

There is will be neither seine nor gillnet fishery in Barkley Sound this year. Toquart Bay is inhabited by ghosts. We have managed our affairs poorly. Even so, there is hope in the air. Maybe the herring are gathering again, somewhere out there, as they have gathered every spring for time out of mind. It's possible. We'll see.

Adult herring from west coast Vancouver Island populations spend their summers feeding on the offshore banks: La Pérouse, Swiftsure and so on. Sexual maturation begins in late summer and throughout the fall, gonads of both sexes grow rapidly. In late fall or early winter, breeding adults migrate inshore, where they remain for several months while their gonads continue to ripen.

In late winter, breeding adults stop feeding to collect in densely packed schools near the spawning grounds. Herring in British Columbia generally spawn between mid-February and mid-April, though some stocks can spawn as early as January and others as late as June. Herring usually spawn in Barkley Sound around the third week of March.

Herring do not form mated pairs; spawning is a group event. When first laid, the eggs are sticky. On Canada's Pacific coast, herring prefer to deposit their eggs on marine vegetation, though some eggs adhere to the rocky bottom. Predictably, the heaviest egg deposition occurs where vegetation is heaviest: the lower intertidal zone and upper sub-tidal. Accumulations rarely exceed eight or nine layers, though much heavier accumulations have been reported. Surviving adults move offshore in scattered, relatively dispersed schools to begin feeding again.

Eggs hatch in about two weeks: 12 to 20 days depending on water temperature. Young herring start life as thin, thread-like larvae, just over half a centimetre long. They grow rapidly during their first six to ten weeks, by which time they have assumed a more fish-like form. Thereafter, until about a year of age, they are called juveniles.

This first period is critical for the strength of the year class: eggs and larvae suffer up to 99 percent mortality. Eggs are lost to predators (mostly birds), desiccation and freezing. For larvae, starvation is the chief source of mortality. Larvae hatch with a supply of yolk but when that supply runs out, they must begin feeding promptly. Within a day or so after yolk sac absorption, the larvae instinctively snap at bite-sized food particles. But if they don't encounter the right kinds of food in bite-sized particles within a day or two, they give up — stop trying — and soon perish. Larval herring are feeble swimmers and subject to a high level of predation from other zooplankton, especially comb-jellies and arrow-worms. Unusually severe pressure from predators or lack of appropriate food at the right time can make the difference between a good reproductive year and a disastrous one.

Juveniles stay near to shore during their first summer, though they may drift from the spawning grounds to nursery areas. During the autumn, juvenile schools drift toward the open ocean; by late October, most have left Barkley Sound.

Juveniles return inshore during their third winter to join the breeding population, merging with returning schools of adults. Pacific herring can live to 15 or 16 years of age, breeding repeatedly — particularly in northern waters — though very few ordinarily survive past 9 or 10 years. Mortality is high and constant for all ages; older age classes dwindle rapidly. In many stocks, particularly those under heavy fishing pressure, the newly recruited juveniles may comprise the largest part of the breeding population and a single poor year class can have a profound effect on stock size.

Timing is everything now as the fish move toward spawning. A day early and there will be nothing to see. A day late and the show will be over. Today, a roaring gale. Tomorrow, a pleasant paddle in the sunshine. I watch the weather reports, read the fishery notices and possess myself in patience.

Even though there is no commercial herring fishery in Barkley Sound this year, the Department of Fisheries and Oceans keeps a vessel and observers on the water to assess the progress and magnitude of spawning. Every day I check their reports, anxious as a fisherman, noting the size and ripeness of the fish, watching for the first sign of spawning.

Finally the announcement I've been waiting for:

Area 23 — Barkley
SPAWN EXTENDS ALL AROUND THE NORTH STOPPER
ISLAND. NO OTHER SPAWN OBSERVED IN BARKLEY SOUND.
Sent March 25, at 11:43:11

By the time I get my gear loaded, the morning is gone. The sky is overcast with showers at Tofino but the road to Toquart Bay is dry. And by the time I reach the boat launch, the sun is trying to break through. At first glance, the place seems much too quiet. My spirits sink: it's a false alarm. Perhaps the DFO observers, ever hopeful, were seeing spawn where none existed.

Then I notice the gulls on an island across the bay. A solid band of white along the high-tide line. Something spooks them and the whole mass erupts into the air, a swirling blizzard of birds. Evidently something is happening over there, but I'm leery of getting my hopes up.

At least it's a pleasant day for a paddle, with just the slightest breeze out of the southwest. The Stopper Islands (so-called, I suppose, because they are the stopper in the broad mouth of Toquart Bay) are a cluster of two main islands, north and south, with numerous satellites. They are farther away than they look and I'm thoroughly out of condition. I rest twice during the crossing, enjoying the sun as it comes and goes.

As I draw nearer to the northernmost island, I begin to understand why the gulls are so nervous. There are bald eagles everywhere. I count a couple of dozen without any trouble at all. They perch quietly in the trees, flap along the shore or soar high overhead on broad fixed wings. Gleaming white heads and tails betray perching birds against the dark green forest background. They seem incredibly tame, but perhaps they're only sated and stupid with food. To my left, a couple of magnificent adults perch on a little rock at the water's edge. They hardly bother to glance my way as I paddle past.

Suddenly an eagle comes off one of the trees on North Stopper Island, flying really hard, stroking along, heading out across the water. What's happening? A gull with a fish, perhaps. The action is too far away, I can't make it out. The eagle swoops to the water, feet and legs lowered. Then it towers into the air, helicopters down to pluck something from the surface and returns to its perch.

I'm two hundred metres away from the north island when a fish jumps from the water just off my starboard bow. A little fish. Then a couple more, some distance ahead. Just now, the sun is shining, finding a break in the clouds. For some reason I glance down and my eyes register a flicker of light a metre or two under the surface. I see another flash and another. In a staggering instant, I realize that the water under the boat is alive, moving, with many hundreds — make that thousands — of little fish, flashing and glinting as they turn periodically onto their sides. It's a little unnerving. Swim and turn. Swim and turn; a slow, almost ritual rhythm, like a procession. The light changes, the vision fades. But at least I know they're down there. The herring have returned. They're not gone for good — not yet anyway — *Deo gratias*. I look up just in time to see a gray whale surface a couple of hundred metres off my port bow. It comes to me that here we have the makings of a very good day indeed.

I stay well out in the deep-water channel until I fetch the farthest northeast corner of the north island. Then I turn toward shore, intending to paddle back over the shallows. The water has an opaque, jade-green appearance. Summer water here. The closer I get, the more impressive the concentration of wildlife. Thayers and mew gulls by the hundreds. Bald eagles. Cormorants. Surf scoters. Porpoises. Seals. Sea lions everywhere — mostly California sea lions, chocolate brown with odd, abruptly bulging foreheads, but also a few Steller sea lions, blond and massive.

Over the next couple of hours, I have sea lions within a boat length or two pretty regularly. This causes me some uneasiness. These are big, powerful animals — up to a ton of muscle, bone and teeth for a Steller bull — completely at home in this alien element. I can't help wondering what might happen if I accidentally get between them and the fish, a question that becomes especially pressing when I blunder into a couple of all-out feeding frenzies. Tremendous splashings and bellowings and brown bodies arching in and out of the water. I visualize shards of fibreglass and tatters of neoprene floating forlornly on the surface.

Fortunately they prove to be surprisingly shy, always giving me a wide berth. On the two or three occasions when a sea lion surfaces anywhere near the kayak, it is gone with a splash as soon as it spots me. Otherwise, alarmed animals bark loudly to alert their fellows. They react oddly: first peering toward me, then surging away as if preparing to dive, then looking back, double-checking, surging away again and so on, huffing noisily the whole time. They have poor eyesight out of the water and it occurs to me that — sitting tall in the kayak — I may look alarmingly like a killer whale with that great dorsal fin projecting high above the body. But not quite. There's just enough of a difference to justify some hesitation and a second look.

For the most part they pay little attention, being much too busy gorging on herring. A big male surfaces not far off, oblivious, with a mouthful of wriggling silver fish, tails still twitching as they swim down his gullet. And then he's gone, back to the larder for another snack.

I'm puzzled as I come closer to the shore. From memories of other trips, I had expected this stretch of water to be quite shallow. But today it shows the dark green of deep water practically to the beach. I'm paddling across that dark green before I realize it's not deep water at all; it's fish. Thousands and thousands of fish, tightly massed. A living river flowing under and around the boat.

This huge school of herring is moving so smoothly and in such concerted unison that individual fish hardly seem to be swimming. Rather it's as if they are being drawn along by some sort of uncanny current, magnetism, gravity, whatever. The spacing and coordination are so perfect that the school changes direction as one organism. It all looks so effortless that I have to think about how amazing this is. All those thousands of fish, complicated manoeuvres, sudden changes of direction. But no collisions. They are not banging into or sideswiping one another. There aren't any pileups or melees, not even any close calls as far as I can see. It's as if each fish were an automaton under the absolute control of some mastermind.

Indeed, the sense of unity is so compelling that I have to work at seeing the individuals. The flow of the school is mesmerizing. But when I make the effort, sure enough, the individuals are there, tails moving, eyes watching, fully alert. There is no master control. Herring are just very, very good at synchronized swimming. It's a skill they have, a talent for coordination.

As I paddle along the shore, the ocean becomes a pale green, the colour of a glacier-fed stream or lake. By the time I reach the island's northernmost point, the water is almost white, like skim milk. This is full-on spawning. Eggs and sperm (milt) are broadcast willy-nilly into the water. This is group mating — a crowd event. For all I know they don't even break formation.

I float at the edge of the milky-white shallows. There is no sign of activity. Perhaps this is just lingering evidence of recent spawning. Have the fish gone back to deeper water? Perhaps this is the spawn that DFO crews noted this morning, not yet dissipated.

As if on cue, a small plane appears, flying low. As it passes noisily overhead, the shallows erupt with writhing fish. The alarm is so intense that some of the spawning herring end up high and dry, grounded. Aerial predators are a major threat; perhaps these fish thought that the great grand-daddy of all eagles had come to make a meal of them. It is a serious piece of disturbance; the whole episode leaves me shaken and angry. In fact, I see a fair amount of human interference over the course of the afternoon. Power boats cut right through the schools, full speed, evidently oblivious. I begin to feel guilty myself. Chastened, I back off and give the area a wide berth. Those fish have already endured enough. At least I know that they are still present and engaged in spawning.

Around the corner the water becomes clear again. But the seaweed, eelgrass and even the bare rocks have undergone a dramatic change in appearance. Every blade and strand is swathed in a greyish mouldy-looking shroud. It isn't until I reach down and pluck up a little piece of kelp that I realize the plants are smothered in layers and layers of herring eggs, the graphic evidence of some recent piscine orgy. Just the few square metres of seafloor under my kayak must support hundreds of thousands of eggs. Up in the sunshine, where I can

see them better, the tiny eggs are transformed into beads of crystal —
the odd one is milky white — about the size of a pinhead. The clusters are extraordinarily beautiful. Try as I might, I see no signs of life.
The little beads seem utterly transparent.

I hesitate a moment before nibbling the egg mass ("When in
Rome ..."). It has an interesting, crunchy texture and a mild fishy
flavour — one of those delicacies that serves best as a vehicle for
interesting sauces, perhaps something flavoured with wasabi.

On shore, another sort of feeding frenzy is in progress: white
seagulls follow the retreating tide down through the golden rockweed, shoulder to shoulder, feathered gleaners, beaks clicking busily
from side to side, little machines gathering the exposed eggs. They
hardly notice me drift past. A wave breaks over and among them.
Taking care of business, they hardly pause.

Farther on, I come across another school of herring, moving in
now-familiar fashion, orderly and smooth. But suddenly, something
seems amiss. The formation is breaking up. The herring, individuals
now, are racing randomly back and forth, through, into and around
the eelgrass. They are — okay, I'll say it — cavorting in the eelgrass,
like cats in catnip. They rub their bodies against the green fronds,
evidently craving the touch of it. They are over on their sides, flashing, glinting, wriggling, undulating through the weed. The whole
scene has an undeniably sensual flavour. Evidently erotic, too. The
water, which had been fairly clear, is suddenly murky. *So, that's how
it starts,* I mutter to myself and paddle on, averting my eyes.

I cross the narrow channel to the south island, past a raucous herd
of sea lions scarfing herring along the shore. It's getting late. The sun
is dipping toward the cloud-capped ridges on the far side of Macoah
Passage. I'm starting to feel the chill. Time to turn back. I hesitate for
a moment, then decide to go just a little farther, to the next point.

In the shallows around the corner, I come into another area of
active spawning, milky white water. But this time, I can see fish down
there. Every now and again, the dim shape of a herring appears in

the milky cloud, as if flung clear of the action. They dart back into whatever riot is happening below, passing quickly from view. I can also see the flashing of fish inside the cloud as they turn onto their sides to deposit eggs or sperm. It's like the flicker of heat lightning in the billowing clouds of a summer storm, a heliograph of ecstasy.

The shallows and sandbars at the mouth of the Maggie River, off to the west, present an incredible spectacle in the golden light of late afternoon. There are thousands of birds out there: gulls, Canada geese, mergansers, mallards, goldeneye, buffleheads, harlequins. I can see a vast, uncountable flock of scoters out in the channel — a packed mass of black birds, drawn into a long windrow. From time to time, part of the windrow lifts into the air, flies a short distance and settles again. It's like the drifting of autumn leaves.

A little farther on, a black bear forages near-sightedly along the shore. Still farther, away out in the sound, a puff of vapour hangs momentarily over the water, like a brief plume of smoke. Another gray whale spouting. I can only imagine the collection of other species, vertebrate and invertebrate, on standby below the surface. All of these creatures, millions of individuals, are waiting for the herring to spawn, waiting for the dinner gong to sound. This is the windfall that kick-starts the biological year on the outer coast of Vancouver Island. Herring eggs and herring bodies are the fuel that will power migration, breeding, egg laying, territorial defence and all the other imperatives of the coming months.

Back home I continue to monitor the fisheries notices. The spawn continues through the next couple of days:

Area 23 — Barkley
SPAWN CONTINUES, AND NOW EXTENDS ALMOST ALL
AROUND THE STOPPER ISLANDS. NO OTHER SPAWN

> OBSERVED IN BARKLEY SOUND, WEATHER IS HAMPERING
> SPAWN OBSERVATIONS.
> sent March 26 at 11:38:20

and

> Area 23 — Barkley
> SPAWN CONTINUES AND NOW EXTENDS ALONG THE VAN-
> COUVER ISLAND SHORE OF MACOAH PASS SEAWARD, AND
> HAS SPREAD TO OTTAWAY AND LARKINS ISLANDS.
> sent March 27 at 11:58:53

Then, quite suddenly, it's over.

> Area 23 — Barkley
> THE MAIN SPAWN APPEARS TO BE OVER IN BARKLEY
> SOUND. MONITORING CONTINUES.
> sent March 28 at 10:16:25

Reproduction in Pacific herring is one of those processes, common in nature, that seem simple enough at first glance but are, in fact, fraught with complexity. The fish move from deep water to shallow water, spawn, then retreat to deep water again. They don't even bother to pair off. What could be more straightforward?

But even superficial examination reveals a web of integration and coordination so elaborate, so elegant, so *ingenious*, that it seems almost consciously designed to give the species its best possible chance of survival. The same superficial examination also reveals the risky nature of herring existence. Life is plentiful but life is cheap. The individual living spark of one egg buried among all those millions has hardly a prayer of making it to maturity. It needs all the help it can get.

It is an article of faith with biologists that a species' life-history characteristics, in general, are not random developments of no particular consequence but are closely related and often of great adaptive significance 'for the success of that species. This is certainly true of Pacific herring. The naturalist, having observed "Who?", "What?" and "When?", is then chiefly interested in working out the adaptive significance of various life-cycle quirks — asking the magic questions: "Why?" and "How?"

Spawning for Pacific herring generally occurs during a period of neap tides midway between the new Moon and full Moon. But specific timing varies according to a complex set of variables. Cold weather, for example, delays spawning. Herring spawn earlier in warm-water years, later in cold-water years. Larger females spawn earlier than smaller females. Southern stocks tend to spawn earlier than northern stocks.

The desirability of avoiding severe weather, with its attendant turbulence in intertidal spawning areas, is obvious. But what benefit might accrue in having temperature affect the maturation rate of females: earlier in warm years, later in cold years? Evidently it sets the timing of egg laying in sync with the springtime development of near-shore ecosystems, so larvae can encounter appropriate food when they hatch. That also explains why herring move inshore so early. They are tuning their bodies to the temperature of the water in which larvae will have to survive.

Water temperature has no effect on the total mass of eggs produced by a single female — ovary size is constant for females of a given size, no matter what the temperature — but females tend to produce more and smaller eggs in warm-water years. Why? Mortality is higher in warm-water years. There is less food for the larvae and more predation. Perhaps a greater number of eggs increases the likelihood that some larvae will beat the odds, even if they are a little malnourished. In cold-weather years, females produce fewer, larger eggs. This results in fewer larvae but they are well endowed to survive and prosper, perhaps a better bet with fewer predators around.

For somewhat analogous reasons, larger females spawn earlier than smaller females. Large females tend to produce larger eggs than small females. Larger eggs, with their greater endowment of yolk, produce larvae better equipped to survive the sparse, early-season growing conditions. Larvae from the smaller eggs of smaller, later-spawning females generally encounter richer growing conditions. Later larvae mature more rapidly than earlier larvae, both cohorts reaching the juvenile stage at about the same time.

Why should herring prefer to spawn on a neap tide? Neap tides are less extreme than spring tides — which occur at the new Moon and full Moon — and the associated tidal currents are less fierce. Slower currents may reduce dilution of milt and improve the rate of fertilization. Less turbulence may also favour the attachment of eggs to the substrate. Perhaps less extreme tides also prevent the herring from inadvertently spawning in the high intertidal, where eggs would be more vulnerable to desiccation, freezing and predation. It might also be that the larvae, hatching two weeks later during another neap tide, are less likely to be swept into deeper, less hospitable waters.

Even group mating, when you think about it, presents some significant challenges. For one thing, individuals have to release eggs or sperm at a precise time without the benefit of stimulation from a mate. Coordination is a real problem. It's hard enough for two individuals in a mated pair to get it together. Imagine a few hundred herring in a mating school. Synchrony is essential. How is it accomplished? Once the fish are ripe, the release of eggs and milt seems to be coordinated by pheromones in the milt. For some reason, a few males release their sperm prematurely. The pheromones in that milt stimulate both females and the remaining males to release eggs and sperm simultaneously. Just one more example of that talent for coordination.

The resulting distribution of eggs over the substrate is surprisingly moderate and uniform, especially considering the frenzy of spawning. Furthermore, herring in successive waves of spawning seem to avoid areas where spawning has already occurred. Both behaviours have great adaptive significance. If accumulations become too great, deeply buried eggs will perish from lack of oxygen or buildup of metabolic wastes. It may be that both males and females test the substrate, avoiding areas that have significant accumulations. Evidence suggests that females may avoid laying eggs on top of other eggs if they have a choice. And it may be that herring are able to detect the intensity of spawning, perhaps through the concentration of milt in the water, with higher levels inhibiting further spawning.

Just the simple fact that Pacific herring choose to lay their eggs on marine vegetation, rather than, say, simply dumping them on the seafloor, is adaptive. The eggs are distributed over a much greater surface area and are suspended in the water column, where the strong currents of the intertidal spawning grounds can promote good healthy circulation in the egg mass. Every little thing helps.

With one thing and another, almost two weeks pass before I can get back to Barkley Sound for a third and final visit. A lovely day, the nicest of all. But even as I drop the kayak into the water, I can't help feeling that the party is over. Still plenty of wildlife out there, but nothing like the extravagant numbers of two weeks ago. Perhaps it's just my imagination, but I sense a general restlessness, too. Everybody is gearing up, getting ready to go. They are like the last few guests after a long banquet, standing on the threshold, exchanging pleasantries, saying goodbyes, jingling their keys. Everywhere I look, all points of the compass, there are little flocks of waterfowl in the air: ducks, geese, the black lines of scoters flying low, fast and purposeful over the water, heading north. The herring themselves, the hosts, are long gone.

I am amazed, even a little disappointed, at how quickly the whole business is winding down, especially when I think about all those months of slow preparation. But I shouldn't be surprised. That is nature's way, from the fading of spring flowers to the falling of autumn leaves. When it's over, it's over; suddenly and irrevocably done, finished, gone. Only human beings, cursed with imagination and memory, are haunted by might-have-beens. Nature hurries on to the next thing, never looking back.

I decide to paddle along the shore toward Macoah Passage. Just inside David Island, I come upon two feeding gray whales, heads down in the water, enormous tails sculling back and forth on the surface. One of them surfaces, swimming toward the open ocean, great barnacled snout breaking waves like the prow of a boat.

The beaches and shallows seem completely clear of eggs, long ago cleaned up by scavenging mouths. But doubtless there are still thousands — millions — left in deeper water. That's what the whales are feeding on. Soon those eggs will be hatching. Already the first wriggling larvae have escaped. I like to imagine what's happening below the surface of the sound, the drama of it. First, just a handful of larvae, then a few hundred, then thousands, then millions breaking free, wave upon wave escaping their little prisons into the cold dark water, vividly alive, to begin the next dangerous phase. The waters thronging with new life. The great spinning flywheel of time moving forward.

Paddling back toward the landing, I catch a puff of warm air wafting from shore, vagrant over the cold ocean, rich with almost-forgotten odours — all those things I miss through the winter — the smell of warm earth, wet wood and budding vegetation. A breath of spring.

> *Wynken and Blynken are two little eyes,*
> *And Nod is a little head,*
> *And the wooden shoe that sailed the skies*
> *Is a wee one's trundle bed.*
> *So shut your eyes while Mother sings*
> *Of wonderful sights that be,*
> *And you shall see the beautiful things*
> *As you rock in the misty sea*
> *Where the old shoe rocked the fishermen three,*
> *Wynken, Blynken, and Nod.*

April

HŌ'UKÅMIL

(Wild Geese Moon)

Springtime at last.

April begins with a break in the weather, a few days of brilliant sunshine, wonderful while it lasts, a false taste of early summer belying weeks of rainy weather still to come. All nature seems on the move. Event follows event in rapid succession; each day brings fresh revelation. Tightly rolled false lily-of-the-valley shoots emerge from the forest floor, unfurling broad green leaves. Catkins adorn the willows. The first ospreys arrive. With herring season winding down, the scoters grow restless. Flock follows flock north-ward along the coast, thousands of birds starting the long journey to northern nesting grounds. Enormous skeins of geese — Canadas, greater white-fronteds, even a few brant if we're lucky — pass overhead on the same errand, talking excitedly to one another and to us down below.

In the old times, once the dried roe was packed away, the clans could move closer to the open ocean so as to be better situated for fishing, hunting and whaling when the calmer weather finally arrived. Men hunted the migrating waterfowl. During stormy weather, birds sheltered in bays and inlets. Using the cover of darkness, two men in a small canoe could approach the resting flocks. On a special platform in the stern, the hunters kindled fire, casting light across the water. The man in the stern handled the boat and manipulated a piece of cedar matting to create a shadow across the front of the boat. The man in the bow handled the net. Bewildered birds, trying to escape the light, swam into the shadow where they could be captured.

We no longer hunt waterfowl in the spring, but those high wavering flocks still have the power to stir our blood when they come calling. Lucky neighbours who've managed to escape to warmer climates for a few weeks or months start drifting back into town, migrating northward like the geese or like the people of an earlier time returning to the outer beaches in anticipation of summer to come.

Hesquiat Harbour

THAR SHE BLOWS!:
The Springtime Migration of Gray Whales

There Leviathan
Hugest of living creatures, on the deep
Stretched like a promontory, sleeps or swims
And seems a moving land, and at his gills
Draws in, and at his trunk spouts out a sea.
— John Milton, *Paradise Lost*, Book 7

Milton may have been a little hazy on cetacean anatomy but he certainly had a tight grip on the one essential fact regarding whales. They're big. Blue whales, the greatest of all great whales — up to 30 metres in length and 136 tonnes — are thought to be the largest animals ever to have lived on Earth, bigger than the biggest dinosaurs. Even gray whales, though not particularly hefty as whales go, are very substantial animals at 15 metres and 30 tonnes — half again as long and twice as heavy as a fully loaded Greyhound bus.

Human beings are fascinated by cetacean superlatives. We love the exotic facts, those aspects of whale biology that make them so different from us. We relish the massiveness of blue whales with their automobile-sized hearts and culvert-sized tracheas. We're fascinated by the thought of sperm whales diving to depths of a thousand metres and staying submerged for an hour at a time while wrestling giant squid in the stygian darkness. We love the idea that great whales may communicate across thousands of kilometres of open ocean with low-frequency sound.

But these *Ripley's Believe It or Not* facts — glamorous though they may be — are essentially superficial. The truly remarkable thing about whales is not that they are so different from us, but that they are so similar. Milton notwithstanding, whales do not have gills nor do they spout seawater — they have lungs and breathe air, like us. The chief cause of death in whales isn't injury, disease or even old age, but drowning. They may be old, they may be badly injured, they may be mortally ill, but drowning is what actually kills them when they can no longer struggle to the surface for another breath. Like us, whales are warm-blooded mammals, maintaining a constant body temperature even in bitter cold water. Like us, they bear their babies in a womb and suckle them with milk. They even have some hair, like us, though the beard on a whale amounts to little more than a few bristles on the top of the snout — whiskers. (I suppose it could also be said, then, that whales, like us, go through life more or less naked. Thirty-tonne skinny-dippers.)

Whales and human beings also share a basic terrestrial pedigree. One hundred million years ago, the distant ancestors of modern cetaceans were little mammals living along the seashore and feeding on small fish, invertebrates and such-like. They must have looked and behaved something like modern-day mink. Over eons, generation upon generation, the little beasts spent more and more time in the water. That shift in habitat must have conferred some benefit on animals willing to take the plunge. Perhaps food was more plentiful in the water, perhaps there were fewer competitors or predators.

On the other hand, spending more time in the water would also have imposed significant costs on those pioneers. The aquatic environment poses a number of rigorous challenges for land mammals. Just getting around is a problem; water is so much thicker than air.

One of the really fascinating aspects of cetacean biology is to consider the solutions that families in this branch of the class *mammalia* have evolved to the problems of life in a fundamentally alien medium. On the problem of getting around, for example: the de-evolution of hind limbs, the modification of front limbs into broad paddles for steering, the development of massive, powerful tails with flattened flukes for propulsion. The kinship between humans and cetaceans lends the subject a special interest; we can empathize. Looking at whales, we see ourselves mirrored back and distorted by deep waters.

Doug Banks is a senior pilot for Tofino Air, a small home-grown air-charter business that operates from the First Street dock. It's his day off and we're chatting in front of his backyard shop. The skeletal frame of a half-assembled Piper Cub is just visible inside the window.

I've come, cap in hand, to ask a favour. Can he offer any suggestions as to how a writer on a budget might go about cadging an airplane ride to Hesquiat Harbour? Two rides, actually — there and back again. I have pressing business at that end of Clayoquot Sound but it's a mighty long trip in a kayak and risky at this time of year with the weather still so unsettled. And I can't afford to charter a plane out-right. Are there other possibilities?

Fortunately Banks shares my interest in natural history. He's sympathetic. "Well," he says, "Pam and John have chartered the Beaver for a family trip to Hesquiat tomorrow morning. They might have an extra seat. It wouldn't hurt to give them a call."

Are the whales still there?

"Oh yes," he says, smiling at some private memory. "The whales are still there."

Every spring, pretty well the entire eastern Pacific population of California gray whales, well over twenty thousand animals, migrates the length of North America's Pacific coast, almost ten thousand kilometres. The adults have just spent three months — December, January, February, perhaps part of March — in the saltwater lagoons of Baja California, calving and mating. Now they're headed for rich summertime feeding grounds in the Bering and Chukchi seas. It is one of the world's truly remarkable natural phenomena, perhaps the greatest mammalian migration on the planet. Imagine an enormous herd of great beasts — every adult bigger than a bus — moving the length of a continent to their summer pasture. It makes the seasonal migrations of the African grasslands seem paltry.

Gray whales are shallow-water whales, bottom-feeders. Unique among cetaceans, they filter small invertebrates from the sand and muck of the seafloor. On migration they hug the coast, more or less, navigating from headland to headland. And every year between late February and early May — but peaking noticeably in late March or early April — the whole procession goes right past my door. These whales are mostly in travel mode, putting the kilometres behind them. For six months they have fasted. Now they are anxious to get back north again and begin feeding in earnest. But occasionally they stop to rest and forage for a few hours in one of the shallow, sheltered waterways along the coast. The herring have just finished spawning and the gray whales are feeding at least partly on roe. In fact, some cetologists believe that the timing of gray whale migration may be determined to some extent by the progression of herring spawn from south to north. The whales exploit this rich but ephemeral food supply to help fuel their journey.

Hesquiat Harbour, at the very northern end of Clayoquot Sound, is one place where gray whales pause to rest and forage, sometimes in significant numbers. I have seen photos of the harbour showing gray whales scattered across its expanse like grazing cattle. But it's a short-lived phenomenon. One day the whales are present in extraordinary numbers, the next day they're gone.

Pam Frazee and John O'Brien are local physicians, husband and wife. When I get Pam on the phone, I discover that they also have whales in mind. They're taking the whole family, Nan and the kids, up for a look. I make my pitch. "Sure," she says. "We should have some extra space. You'd be welcome."

We meet next morning at the dock. It's a working day on the waterfront with lots of traffic coming and going across the harbour. The weather is fine, a bit of luck at this time of year. The big float-plane looks very spiffy in the sunshine — freshly painted in Tofino Air's colours, green and white, with the company's mascot, a seagull in slicker and sou'wester, prominent on the aircraft's tail. I'm delighted to discover that Doug Banks himself will be taking us out. I'm first aboard to take my place at the rear of the aircraft. Nan and the kids are next, then Pam and John up front. Finally Banks climbs in, runs his checks and fires up the big rotary engine. The dock manager unties our mooring lines and the Beaver eases into the ebb tide.

We motor upstream until we've got sufficient room to taxi. Then Banks brings the plane around and sets her on a line for Stubbs Island. He eases the throttle forward and the aircraft gathers speed rapidly, bouncing over the chop, skimming the surface of the harbour. The noise is deafening. When the airspeed indicator touches 60 miles per hour, he coaxes the plane up onto one float and then takes us into the air. The Beaver clears the trees on Stubbs Island and climbs west-ward across Father Charles Channel. A group of kayakers make their way along the southeast coast of Vargas Island. Paddle blades flash in the sunshine, a measured rhythmic semaphore, message unknown.

Above Ahous Bay, we bank right to follow the coast north. Far below, a whale feeding in the shallows lies half-hidden in the plume of muddy water it has stirred from the bottom. Half a dozen tour boats, bright points of synthetic yellow and red, vivid against the blue expanse of water, attend Leviathan at his supper.

One of the problems that whales have, being big, is finding sufficient sustenance. It takes a powerful mess of food to sustain those tons of flesh and keep their metabolic fires stoked. All whales are predators but they fall into two distinct categories with very different foraging strategies. Toothed whales — orcas, sperm whales, dolphins — feed on relatively large prey that they actively pursue, seize with simple pointed teeth and gulp more or less whole.

Baleen whales — gray whales, bowhead whales, blue whales — feed on relatively tiny prey, small fish and crustaceans. It seems ironic that Earth's largest predators should hunt such tiny creatures. But it makes sense for very large animals to look for sustenance at the bottom of the trophic pyramid, where a greater mass of food can be easily had, than at the top, where individual items of prey, though large, are scarce and hard to catch.

How do such large animals capture and hold such tiny prey? And, more subtly, how do they manage to eat that prey without ingesting a lot of very salty ocean as well? In short, they've managed to evolve a sort of colander in their mouths. They have given up teeth. Their gums develop a growth of baleen plates, long sheets of keratin — the protein of fingernails — hanging vertically in their cavernous mouths and overlapping like the slats of a Venetian blind. The inner edge of each plate is frayed into fibres or bristles. The mat of bristles inside the overlapping plates serves as an effective net or strainer. The size of the mesh thus formed varies between different species of whale, depending on their habitual food: coarser for larger prey, finer for smaller prey. When baleen whales have a mouthful, they use their tongues to press the seawater out of their food. The water escapes through the baleen while the food remains trapped inside.

Baleen whales can be further sorted into three categories according to differences in their approach to the same basic feeding strategy. Rorquals — blues, fins, humpbacks — are gulpers. They have large mouths and deep pelican-like throat pouches. When rorquals encounter a school of prey, they open wide and engulf the whole mass. Right whales and bowhead whales are continuous strainers. They swim through the ocean, mouths wide, living plankton nets, straining food through long baleen plates as they go. Gray whales, in a class by themselves, are bottom-feeders. A gray whale lays the side of its head on the bottom and sucks up a giant-sized mouthful of sand and mud containing ghost shrimp, amphipods and other prey. The whale chews to loosen the mass, then uses its tongue to squeeze the soupy mud back into the ocean through the baleen, leaving only the prey behind to be swallowed.

The green slopes of Meares Island and Vancouver Island rise sharply from the sea on our starboard side. The brilliant morning sunshine casts the folds of the landscape into high relief, exaggerating the ruggedness of the countryside. Long, narrow inlets follow steep-walled glacial valleys back into the heart of the high country that forms the spine of Vancouver Island. In the more sheltered water-ways, meadows of eelgrass — a much brighter green than the forested islands — spread across the shallows and sandbars.

On the other side of the aircraft, the wrinkled ocean, glinting in the sun, reaches to the western horizon. Out along the edge of the conti-nental shelf an oil tanker bound for the Strait of Juan de Fuca drapes a plume of smoke between sea and sky. Below the floatplane, where land and ocean meet, lies a fractured archipelago of minor coastal islands, densely forested and laced with kilometres of sandy beach and rocky shoreline. The whole length of the exposed coast is white with surf. It's a good thing I'm not down there trying to paddle to Hesquiat.

We cross Russell and Brabant channels and fly on past the moun-tains and sweeping beaches of Flores Island, past Sydney Inlet, past Hot Springs Cove. Finally, up ahead, I can see the great low expanse of the Hesquiat Peninsula curving out into the Pacific. Even at this distance I catch the gleam of the Estevan Point light station — though its beautiful tower, supported by graceful flying buttresses like some medieval cathedral, is too far away to make out. The arc of the penin-sula encloses Hesquiat Harbour, a broad but relatively shallow bay, the northernmost end of Clayoquot Sound.

In many ways, Hesquiat Harbour is the ultimate expression of the natural-world vividness that makes Clayoquot Sound such a special place. The harbour is exceptionally rich in marine life: shellfish, fin fish, birds and mammals. The shoreline comprises a full range of habitat — rugged rocky sections, large sea caves, sandstone shelves, sand, pebble and boulder beaches — all supporting rich intertidal communities. There is a human presence, certainly, including traces of ancient engineering, boulders moved for canoe paths, fish weirs and fish traps, but disturbance is limited and the overall impression is of a wild landscape.

I wish I could say the same for the surrounding countryside. Wherever I look, whichever way I turn, clearcuts cover the rugged mountains, up one slope and down the next. It's an appalling display, as bad as anything in Barkley Sound. Nor is the damage confined to a loss of forest cover. There are landslides everywhere. Great piles of debris choke the mouths of blown-out streams and blanket once-productive intertidal areas. And there's more grief to come: those steep slopes have barely begun to stabilize.

Banks throttles back and the aircraft begins to lose altitude. We follow the shoreline, looking for whales. On the east side of the harbour, we overfly a sea lion haul-out, a couple of enormous rock slabs, wave swept, covered by a mass of chocolate and golden-brown bodies. Like the whales, they are here because of the herring. And I can see places along the shore where the water is still milky with the residue of recent spawning.

But where are the whales? Two or three minutes pass and still no sign. Discouragement begins to set in. Have they moved on? Then Banks shouts: "There's one." He points ahead and to the left. From the rear of the plane I can't see anything, but we're banking sharply and I find myself with an excellent view, straight down, of the jade-green water below the aircraft. Suddenly, before my very eyes, two massive shapes materialize from the murky depths, break the surface, blow plumes of water vapour — the condensation of their exhaled breath — and dive again. The ocean is so thick with plankton that they're gone from view almost immediately.

Large whales probably cannot see their own tails much of the time. In the clearest tropical waters, three-quarters of available sunlight is lost in the first ten metres. The ocean is pitch dark at depths greater than 200 metres. In the murk of cold plankton-rich waters along the west coast of Vancouver Island, visibility is even more limited: sunlight penetrates a mere 15 to 35 metres.

Whales have compensated by evolving exceptionally acute hearing. The process has been less straightforward than you might think. It wasn't just a matter of growing larger, more sensitive ears like a fox or a bat. Mammalian ears evolved to work in air. Submerged in water, we lose our capacity to determine the direction of a sound. Carried through the liquid and bone of our bodies, rather than our external ears, a sound seems to come from everywhere at once — try it in the bathtub sometime.

So cetacean evolution had to substantially rework the anatomy of the skull and inner ear. The bones enclosing the middle ear, firmly attached to the rest of the skull in other mammals, float free in whales, anchored only with soft tissue. The bones inside the middle ear have been modified to capture vibrations from areas along the sides of the head, rather than from the external ear, and the whole structure is surrounded by pockets containing a mucus foam that provides acoustic insulation.

As a result, whales have excellent underwater directional hearing. The toothed whales can also generate precise high-frequency sound pulses, and evidence suggests that their brains are good at interpreting the returning echoes. In effect, they are equipped with biological sonar, active echolocation. Cetologists are not certain whether baleen whales employ active echolocation; they may simply echolocate passively.

We circle the head of the harbour and come in low, just skimming the surface. The aluminum floats touch down, skidding. The plane slows rapidly and settles into the water. We idle forward, edging closer to the place where the first whale blew. Banks eases off on the throttle and kills the engine. We rock quietly on Hesquiat Harbour's broad breast, enjoying the sunshine, waiting for events to come our way.

A whale spouts in the distance, then another. Once we have an idea of what to look for, we start to see whales spouting all over the harbour. It's a lovely moment, peaceful and quiet. The whales go about their business while we sit watching and talking quietly. Banks opens the doors so we can hear as well as see. Even the kids are quiet and content.

The sun seems exceptionally bright and warm for the time of year. No surprise, I suppose. It's unusually calm today, here in the middle of the bay, hardly a breath of wind to cool us. And of course you get a double helping of sun when you're on the water. The surface of the harbour is a liquid mirror, undulating gently. Water laps against the Beaver's floats, a hollow metallic slap. Scoters commute back and forth.

Such experiences in the natural world are much sought after nowadays. They have even become a commodity of sorts. Organized whale-watching tours, which simply didn't exist fifteen or twenty years ago, are now a major recreational activity and a thriving industry has grown up to meet the demand.

There is something about whales — their size, their intelligence, their beauty, their oddness — that seems to capture the popular imag-ination. Whale watchers in Clayoquot Sound might see killer whales, humpback whales or, very rarely, something more exotic like a minke or a sperm whale. But gray whales are by far the most common large cetaceans and they are the stock-in-trade of the local whale-watching industry.

Every spring, the communities of Tofino and Ucluelet, in cooper-ation with Pacific Rim National Park, stage the Pacific Rim Whale Festival to celebrate the northward migration of gray whales. Thousands of people journey to the outer coast of Vancouver Island hoping for a glimpse of the whales. For two weeks a committee of local volunteers serves up a variety of entertaining and educational activities, a curious mix of tourism-industry hype and genuine affection for the animals.

For visitors who can't make it in the spring, local charter operators continue to run trips through the summer. Fortunately there are always a few whales that decide to cut short their long trip to the Bering Sea. Why bother swimming all that way when there's plenty to eat right here? These so-called resident grays can be observed into the early fall, feeding or loafing in shallow bays and inlets across the mouth of Clayoquot Sound.

When half a dozen whale-watching boats descend upon a couple of feeding whales, it can be a bit of a circus. Given the high capital costs, the attractive potential profits and ever-increasing competition for paying customers, operators are under considerable pressure to provide their clients with a worthwhile view. There is real concern among the charter operators themselves that increased traffic may somehow be compromising the animals. But for the present, research suggests gray whales are not particularly disturbed by all the attention. There are no obvious signs of distress and the animals show little or no change in feeding behaviour when the tourists arrive. The whales may spend slightly more time on the surface when boats are present. Indeed, some whales seem to relish the attention. These "friendly whales" occasionally seek out boats and approach them closely.

Even so, whale watching is a pretty mild entertainment, generally speaking. It's a rather different experience from watching terrestrial wildlife. When you watch a bear, a moose, a hummingbird, you can observe the animal moment to moment, going about its business. You get a clear sense of that animal's life; you can participate in a small way. With whales, the best you can hope for is a brief glimpse. Most of the animal's life is mysterious, obscure, hidden. All you can take away is an awareness of the creature's presence and perhaps some sense of its life as a function of the environment it inhabits.

The chief reward for getting out to watch the springtime migration of gray whales is the sense one gets of being witness to a stupendous natural event. Fossil evidence suggests that gray whales have been migrating up and down the west coast of North America for at least fifty thousand years, perhaps much longer. During every one of those fifty thousand springs, gray whales paused in this very place to refresh themselves, though there must have been intervals when they had to dodge glacial icebergs. And in coming out to watch, we play our own modest role in that story, that great timeless pageant, if only as observers.

Suddenly a whale surfaces right beside us. The animal is close enough that I can see its eye, surprisingly small for such a huge body. The eye is watching us. The plane bobs in the spreading ripple. The whale's skin is a mottled grey encrusted with barnacles. Perhaps because they are slower moving or because they spend so much time grubbing around on the bottom, gray whales accumulate a load of external parasites, including one highly specialized species of barnacle that lives nowhere else.

The whale exhales quickly and then inhales, just as quickly. I listen through the open door. I smell the fetid odour, see the little rainbow in the droplets of exhalation. The sound of a whale breathing is not what I expected. You might think that this large ponderous animal should take large ponderous breaths. Not so. Our whale gasped for air — that's the only word to describe it — a very brief, sharp exhalation followed immediately by a quick inhalation. Think of a small child paddling in cold water, mostly holding his breath, but every now and then gasping for air, while trying not to inhale any water. That is the sound of a whale breathing and it is a revelation to me, an unexpected point of kinship. A minor epiphany. I take away a stronger sense of empathy for this animal. It's one of us.

The whale arches its body downward and dives in a leisurely way, sinking from view. For a moment the great tail — three or four metres across the flukes — rises clear of the surface, streaming water. It hangs there for a moment, then follows the whale down into the ocean and out of sight. Whale watching doesn't get more vivid than that.

Now here's a curious thing. When our whale surfaced there was a burst of astonishment and excited chatter in the plane, but no alarm. It did not seem to occur to us that we might be in danger, though the creature could easily have swatted us out of existence. Perhaps we were being naïve but I detected no menace down there. Nor fear, for that matter.

There is a good deal of irony in that, considering the former hostilities between our two species. When whales were still hunted from small wooden boats, gray whales had the reputation of being particularly dangerous. Hunted cows defended their calves fiercely, attacking the boats. Even so, California gray whales have twice been hunted to the brink of extinction. Happily, the hunting of gray whales was banned in 1939 and the eastern Pacific population has recovered almost its full strength. It's one of the very few success stories in cetacean conservation. Atlantic grays, by comparison, have been completely extirpated and populations in the western Pacific remain severely reduced.

We wait for our whale to reappear, hardly daring to breathe. After a few minutes with no further action we begin to relax, whispering, comparing notes. Eventually the whale — or we think it's the same whale, who can tell for sure? — surfaces about a hundred metres away. Thereafter, though we continue to watch, it becomes impossible to identify "our" whale with any certainty. It is simply one of the herd feeding out in the harbour.

Clearly the most pressing problem for any former land mammal now spending most of its time submerged in seawater is to get and carry sufficient oxygen. Whales have not managed to evolve gills. They still breathe air and are absolutely tied to the surface. But their food is underwater, sometimes more than a kilometre deep. They must remain submerged for extended periods. How do they manage it?

For one thing, cetaceans are extraordinarily well equipped for breathing. They can empty the air from their lungs and inhale a fresh supply more quickly and efficiently than other mammals. They have large-diameter airways: the windpipe of a blue whale is big enough for a man to crawl through. They have broad strong diaphragms, the sheet-like muscle that draws air into the lungs. They have exceptionally elastic lungs to squeeze out the stale air. They can empty and refill their lungs more completely than land mammals, typically exchanging up to 90 percent of their lung capacity on each breath; human beings use 25 percent or less.

Cetaceans are also very efficient at extracting oxygen from the air in their lungs. Cetacean corpuscles — red blood cells — are extra-large, extra-numerous and loaded with hemoglobin, the protein that captures oxygen from the lungs and carries it throughout the circulatory system.

Capturing oxygen is one thing, storing it quite another. The massive muscles of a whale are loaded with a hemoglobin-like protein called myoglobin. Cetacean muscles contain up to eight times the myoglobin of human muscle, weight for weight. Myoglobin has a much stronger affinity for oxygen than hemoglobin. Just as hemoglobin captures oxygen molecules from a lungful of air, myoglobin strips those oxygen molecules from the hemoglobin. The muscles themselves become tremendous reservoirs of oxygen.

The point is that whales don't store oxygen in their lungs. They are not "holding their breath" during a dive. They store oxygen in the hemoglobin of blood and the myoglobin of muscle. That's important because when whales dive, dramatic changes occur in both the respiratory and circulatory systems. At depths greater than 100 metres, pressure collapses the alveoli — the little membranous sacs in mammalian lungs where gaseous exchange actually occurs. Air is forced into the bronchi and windpipe, where the remaining oxygen is unavailable to the animal. At the same time, as part of a "diving reflex," the flow of blood is restricted to all parts of the body except the heart and brain, two critical organs with a high, steady demand for oxygen and very little storage capacity. Because the heart is circulating a lower volume of blood, its workload is considerably reduced, diminishing the demand for oxygen. The heartbeat slows, further diminishing demand. All this serves to extend the length of time the animal can stay underwater. But it leaves muscles out of the loop, fending for themselves; that's why they must carry their own oxygen supply.

A whale's lungs represent a surprisingly small proportion of body mass: three percent, as compared to seven percent for humans. That might seem illogical. Surely an animal that "holds its breath" for extended periods should have very large lungs. But oxygen is quickly stripped from the air in the lungs to be stored in blood and muscle. Air in the respiratory system isn't available to the animal, especially on a deep dive. Further, large lungs would be a tremendous handicap to a diving animal, like wearing an oversized life jacket. The additional buoyancy would be a terrible problem.

Why don't whales get the bends? A scuba diver working at depths greater than 10 metres must follow strict decompression procedures when returning to the surface. Nitrogen that has dissolved in the diver's blood under high pressure during the dive must blow off little by little, breath by breath, as the diver returns to the surface. This takes time. The deeper and longer the dive, the more nitrogen has dissolved in the blood and the more prolonged the ascent.

If decompression procedures are not followed, if the diver returns too rapidly to the surface, bubbles of nitrogen come out of solution in the blood — like bubbles in a bottle of soda water when the cap is suddenly removed. The fine bubbles of nitrogen block capillary circulation and the diver suffers from decompression sickness, "the bends," with symptoms ranging from mild confusion and fatigue to chest pains, shortness of breath, paralysis, convulsions and death.

But sperm whales can return abruptly from a thousand-metre dive, swimming straight to the surface. How do they manage it? The answer is simple. Unlike human scuba divers, whales aren't breathing down there. On deeper dives their alveoli collapse and they can't access the air in their respiratory system. They aren't absorbing any high-pressure nitrogen on a dive, so they have none to blow off.

Time to go. We close the doors. Banks engages the starter, cranks the engine two or three times. It coughs and roars into life, thank goodness. We taxi toward open water, turn and accelerate. The Beaver cuts a white slash across the calm blue expanse of the harbour and lifts itself into the sky, carrying us back toward Tofino and home.

I find myself face to face with the ravaged mountains again. Plenty of food for thought there. Consider, for instance, the fundamental similarities between the "harvesting" of old-growth forest and the "harvesting" of gray whales. Until very recently of course, whales, like trees, were there to be expeditiously slain and converted to commercial product. The only useful whale was a dead whale. To that way of thinking, the entire natural world represented a wealth of raw material and potential profit. Even birds, passenger pigeons, Eskimo curlews, golden plovers, were treated as prime items of public nourishment and perfectly legitimate commercial products — commodities. They were exploited with the same single-minded focus on resource extraction, the same avarice that I see displayed so grossly on the slopes around Hesquiat Harbour — the same devotion to profit that almost extinguished gray whales.

Now here is an interesting thing. At some point, perhaps when they became too rare to excite greed, small birds ceased to be objects of ruthless exploitation and became, instead, objects of concern, enthusiasm and — ultimately — affection. City folks no longer depend for their supper on somebody going out to nail a blackbird or a curlew. Most urban consumers would be horrified by the thought.

Commercial exploitation of whales continued much longer — and cetaceans are still taken in some parts of the world. But whales now seem well on the way to making that same transition from resource object to object of affection.

There seems to be a pattern. First comes ruthless exploitation, a feeding frenzy, a reckless strip-mining of wealth from nature. There is a disregard for the future of ecosystems and individual species that borders on contempt. The very possibility that there might be any problem is denied, often vehemently. Then, at the eleventh hour, a change of heart, a sudden panicky concern for the survival of whatever entity is now teetering on the edge of extinction. And thereafter, perhaps, a genuine interest and affection — a form of emancipation really — too late to restore the object of affection to anything like its original abundance, but maybe sufficient to keep it from disappearing altogether.

The obvious question raised by the clearcuts outside my window is this: When will that same sort of emancipation be extended to forests? When will it become unthinkable to kill a thousand-year-old cedar? Is there some way to short-circuit the process and jump-start our commercial culture's sluggish sense of ethics toward the natural world?

Back at Hesquiat Harbour, the drone of the aircraft's engine is already fading, leaving a profound stillness. The Beaver shrinks, becomes a tiny dot against the mountains and is gone from view. The whales continue feeding as before.

May

INIHITCKMIL
(Getting Ready [for Whaling] Moon)

In May, spring comes busting out all over. By month's end, the alders are covered with fresh green foliage. The dogwoods are in bloom along the highway. On the beach, Indian paintbrushes are blooming; ditto beach peas and wild strawberries. The salal is blooming, the evergreen huckleberries are blooming, the bunchberries are blooming, the buttercups are blooming — everything is blooming. The birds sing early and late. The dune grass has already set seed and the first salmonberries are ripe. Summer now seems just around the corner. But winter hasn't quite finished with us yet. The weather turns cold and wet for late May and June.

In the old times, only chiefs hunted whales, beginning in late spring; an act of great physical courage. A crew of eight men — the chief, six paddlers and a steersman — might spend days in their open canoe, weathering storms and heavy seas, sometimes venturing far from shore. And there was always the danger of being stove in by a wounded whale. Humpbacks were preferred, richer in oil and more buoyant, but gray whales were also taken. The barbed, detachable head of the harpoon was fixed to a cable attached to a series of inflated animal skins. The wounded animal dragged this gear, exhausting itself, until it could be approached and killed with a lance. Just getting the dead whale home was a major undertaking. A crew might have to paddle for many days, towing the carcass, particularly if the whale had run out to sea. Ultimately the carcass was beached and flensed, then distributed according to strict protocol. Individuals had rights to particular portions depending on their status in the community or the role they had played in the hunt. Blubber was more valuable than the meat, which had often spoiled by the time the animal was finally processed. Blubber kept much longer. It could be eaten directly or rendered for oil. Just about everything was eaten dipped in whale oil if there was any to be had; a plentiful supply of oil was a respected symbol of wealth.

We no longer hunt whales. The best we can claim is the heroic act of waiting patiently: waiting for the weather to break for good, waiting for school to be over, waiting for summer to come. The town's population swells steadily toward its summertime high. College and university students, free at last, come west to pursue summertime jobs, summertime romance, maybe a little summertime surfing. But real summer is still a month or more away.

Parry Passage

Esowista Peninsula

WIND BIRDS: Mudflats and Shorebird Migration

*The restlessness of shorebirds, their kinship with the dis-
tance and swift seasons, the wistful signal of their voices
down the long coastlines of the world make them, for me,
the most affecting of wild creatures. I think of them as
birds of wind, as "wind birds."*

— Peter Matthiessen, *The Wind Birds*

Early May. I force my way through dense brush, following a faint path
downhill toward the Browning Passage mudflats. My weather-luck is
holding: I'm blessed with another lovely day. Sunlight filters through
the forest canopy to illuminate the exuberant vegetation crowding
along both sides of the trail. The blue sky is speckled with little clouds
drifting out of the northwest.

I have come for a look at one of Clayoquot Sound's most remarkable — and least heralded — natural events. Even before I come to the
edge of the forest, I can hear the subjects of my quest, somewhere
beyond the screen of trees, talking, singing, quarrelling. It's an odd,
interesting sound, a faint musical twittering like the distant tinkle of
wind chimes — or, as W. H. Hudson wrote: "like the vibrating crystal
chiming sounds of a handful of pebbles thrown upon and bounding
and glissading musically over a wide sheet of ice."

It's the sound of very many small birds all vocalizing at once.

Over a period of weeks, in late April and early May, whenever the
sky clears and the wind comes round to the northwest, great flocks of
shorebirds settle onto the mudflats of Tofino Inlet to await the next

southeasterly gale. There are whimbrels and godwits. There are dun-lins and black-bellied plovers. There are dowitchers and sanderlings and knots. Most especially, there are thousands of tiny western sand-pipers keeping company with even tinier least sandpipers. Every single bird is in the midst of its own incredible journey. Already they have come a great distance. Many of them spent the winter — our winter — south of the Equator and most will travel all the way to the high Arctic for a brief breeding season.

This impressive spectacle goes largely unnoticed. When herring gather to spawn, when gray whales arrive on migration, when salmon are running, people sit up and pay attention. But every spring, shore-birds appear by the tens of thousands, spend a few nights and move on without arousing more than a ripple of interest outside a small circle of aficionados.

The general indifference is even more surprising when you con-sider that this spectacle takes place not in some remote corner of the sound but practically on our doorstep. I didn't even need to get my truck out of the shed this morning. An easy walk, less than half an hour from home, has brought me to this place. Even so, it remains the path less travelled: I have yet to see another human soul.

A little geography. The village of Tofino lies at the very tip of the Esowista Peninsula, which extends northward from the low country west of Kennedy Lake and includes the Long Beach Unit of Pacific Rim National Park. The peninsula is almost 17 kilometres long, but barely more than a couple of hundred metres wide in places. Few of the area's visitors, driving the highway back and forth through the green tunnel of forest, realize just how close the ocean presses on both sides.

To the west lies the open Pacific. The outside of the peninsula sup-ports the broad, sandy beaches, surf-swept and restless, for which the area is justly famous: Long Beach, Schooner Cove, Cox Beach, Chesterman Beach and all the rest.

To the east, Browning Passage cuts between the peninsula and Meares Island. It's a very different sort of environment on that side, much calmer than the open coast, protected from wind and surf, sunnier, warmer. On summer days especially, when the outer coast is shrouded in fog, the warmth radiating from the peninsula's forest is enough to evaporate mist drifting in from the sea. The air over the inlet is clear, the sun bright and hot. The dark expanse of mudflat acts as an enormous solar trap; temperatures can be many degrees higher than on outside beaches. On a rising tide, seawater flowing across the warm mud becomes almost tepid. A good deal of fresh water runoff mixes with the surface layers of seawater.

Wherever significant amounts of sediment-laden fresh water mix with salty, nutrient-rich seawater, a series of chemical and physical interactions cause the suspended material to precipitate in minute particles. Strong tidal currents, sweeping back and forth, keep the main channels of Browning Passage and Tofino Inlet clear. But in sheltered bays where the current is not so strong, particles settle out of the water. Over thousands of years this process has given us a series of enormous mudflats — Arakun Flats, Ducking Flats, Doug Banks Flats, Maltby Slough, South Bay Flats, Grice Bay Flats — mirroring the sandy beaches along the outside of the peninsula.

Mudflats belong to a particularly fascinating category of intertidal ecosystem: the estuary. Estuaries take many forms — mudflats, salt marshes, sloughs, eelgrass meadows — but as a group they comprise some of earth's most productive and useful environments. More sheltered than most intertidal environments and richer in nutrients, estuaries support extraordinary communities of plants and animals.

Central to the economy of these communities is a phenomenon called the detritus cycle. Decomposition of organic material releases minute particles of insoluble residue into the water. Drifting about, these particles of detritus become coated with organic molecules and bacteria.

The coated particles serve as food for filter-feeding animals, clams and such, and for substrate ingestors like annelid worms. The animals digest what they can and excrete the insoluble particles to begin the cycle anew. The detritus-feeders themselves fall prey to predators, which are taken in turn by larger predators. And so it goes, all the way up through the trophic pyramid.

Life isn't all a bowl of cherries for estuarine animals and plants; it's not an easy place to live. Aside from the ever-present risk of becoming someone else's breakfast, estuarine organisms must cope with tremendous fluctuations in oxygen levels, salinity and temperature. Even so, a great many species depend upon estuaries at some point in their life cycle. The mudflats of Browning Passage and Tofino Inlet abound with small fish and invertebrates. The mud is honeycombed with burrows. It's a fascinating place to explore, full of bizarre life forms hidden beneath rocks and buried in the mud. Many of those life forms rate highly on the shorebird menu. To a flock of sandpipers cruising past at an altitude of three or four thousand metres, these big mudflats must look well-nigh irresistible.

Where my trail breaks from the forest, I emerge onto a narrow meadow of grass and sedge along the edge of the mudflats. The few minutes I spent checking tide charts earlier this morning are proving to be an excellent investment. The tide is at maximum flood; my timing is perfect. The inlet is full to the brim and seawater laps gently against the fringes of shoreline vegetation. The mudflats are completely submerged, invisible except for a narrow band of brown ooze along the shore.

Crowded onto that narrow band of mud and vegetation is a packed mass of birds, hundreds and hundreds of individuals, perhaps a dozen different species. The noise is louder here, not unpleasant but steady and insistent. It's a comfortable, conversational sound. The feathered mass shimmers and flickers with constant movement. Birds preen and stretch. Birds quarrel with their neighbours. Birds forage in the mud for food. It's an impressive sight, a vivid mosaic of colour and sound.

A hundred metres farther north, the shoreline doubles back into a little bight. A small creek flows from the forest into the sea. The sheltered spit thus formed is one of the best pieces of shorebird habitat in Clayoquot Sound. Moving in that direction, I keep well back from the water, anxious to avoid disturbing the flock.

I needn't have worried. The birds seem to accept my slow-moving presence quite calmly. Perhaps, as far as they're concerned, I'm just a different sort of bear going about my business, which doesn't include eating shorebirds. As it happens, a big black bear comes through here quite regularly; I've met him. These birds probably see that bear every day. I'm just bear number two.

Since I've made my approach in full view, there seems little point in trying to hide now. I pick out a comfortable piece of driftwood and sit down. The birds are so close I hardly need to bother with binoculars. This is wildlife viewing at its most convenient. I watch them; they watch me; everybody's happy. It's an odd setup for bird-watching and I feel almost indecently exposed. But the feathered mass along the shore seems more or less oblivious.

And what a pleasant spot it is. The air is calm. The sun is warm on my shoulders. I've brought a pad along to provide a little insulation from the cold ground, usually most welcome but hardly necessary today.

Browning Passage is at least a couple of kilometres wide at this point, a broad panorama. Across the water, the forested slopes of Mount Colnett and Meares Island rise steeply from tidewater. On my left, some distance away, lies the little archipelago of islands at the mouth of Lemmens Inlet that I explored in February. Beyond them, in the distance, the snow-clad peaks of central Vancouver Island — Mount Mariner and all the others — soaring into the blue.

Even from this distance I can see that the deep blue water on the far side of Browning Passage is spangled with whitecaps. A whale-watching boat, returning from a trip to Grice Bay, works its way north along the channel, heading back to Tofino. They're having a rough ride, those folks. The boat is pounding into the chop, throwing up spindrift, clouds of spray. Pretty windy out there, a regular gale out of the northwest. But here, in the sheltered lee of the peninsula, there's barely enough breeze to stir the grass. These birds, seasoned travellers all, know where to stop for a decent meal and good night's rest.

At such close range, the feathered mass resolves itself into individual creatures, even without benefit of magnification. When I train ten-power binoculars on them, the finest details come into focus: legs, beaks, feathers, bright little eyes watching me. *That's better*, I think. That's the way to think of these birds, an important shift in perspective, from flock to individual.

A flock of shorebirds is an extraordinarily tight unit. Anyone who has seen these birds in flight can attest to that. The semaphore flash of wings as hundreds of birds turn in unison, turn and turn, seems almost supernatural. They remind me of the herring schools in Barkley Sound: such perfect coordination. As with the herring, it would be easy to presume that individual birds are little more than automatons programmed by instinct, simple mechanical subunits of a larger machine.

But the close-up view dispels that fallacy. At ease, these birds are as individual as any crowd of tourists: this one is fussing over its toilet, that one is sleeping, yet another is chivvying a neighbour. Doubtless their behaviour is shaped largely by instinct — whatever that means — and they do have a strong urge to conform, but shorebirds are not machines. They are conscious of the world around them, processing experience and reacting appropriately, just as human beings do. I have no doubt of their intelligence. Even so, I need to work at taking the next step: not merely thinking of each creature as a self-determined unit governing itself according to the needs of the moment — without, I might add, any benefit of rules, regulations, policies and procedures — but as an individual with a life and a history.

A couple of weeks ago, that western sandpiper over there, the one stretching its wings, would have been enjoying a tropical beach — in the Caribbean, perhaps, or somewhere along the coast of Peru or Ecuador. It knows what the Pacific coast of North America looks like from five thousand metres. In a couple of weeks it will have chosen some spot on the tundra in northwest Alaska and be well into the process of creating the next generation of western sandpipers. In short, that little bird leads a more interesting and adventuresome life than I do. Who am I to dismiss it as an automaton, a bundle of reflexes?

Too often, ornithology deals strictly in numbers and occurrences. Birds are treated as little more than vivid objects to be counted, described, tabulated. But each bird is an individual like you and me with its own curriculum vitae: challenges met, successes achieved and calamities suffered. It is interesting to think about what each bird has achieved and what the costs of that achievement might be. The idea of an "average bird" is fiction. Each individual has its own unique life. There are successful birds and there are birds that — for whatever reason, simple bad luck, perhaps — have a difficult time.

Our understanding of shorebirds has grown tremendously in the last fifteen or twenty years. Newer techniques, including radio and satellite telemetry, have revealed amazing things.

These little western sandpipers, for instance, may be flying from Peru to Siberia, muscle power all the way, covering up to 1,800 kilometres a day, sometimes staying aloft a day and a half at a time, flying at altitudes up to 5.5 kilometres. But among shorebirds, they have nothing much to brag about. Sanderlings, not much larger than the westerns, may have come all the way from Tierra del Fuego. Ditto semipalmated plovers. Bristle-thighed curlews migrate from Alaska to Polynesia or New Zealand. Wandering tattlers breed in Alaska and winter as far west as northern New Zealand and the Great Barrier Reef of Australia. Pacific golden plovers can migrate 4,500 kilometres non-stop; those breeding in western Alaska or Siberia may winter as far away as Africa.

When I pull into Shirley Langer's driveway next morning, the sky is bright overcast. The air is still, but I've a feeling the wind is coming around from northwest to southeast. A change in the weather is on the way. In fact it's just starting to rain: a few scattered drops on the windshield.

Shirley is halfway down the stairs when I reach the house, an older woman, cheerful, binoculars in hand, wearing raincoat and rain pants. She had expressed interest in seeing the birds and we made the arrangements last evening. Adrian Dorst, Tofino's resident ornithologist, will be meeting us at the mudflats.

She laughs as she settles into the passenger seat. "Got your rain gear?"

I nod. "Never leave home without it."

Adrian is just getting out of his van when we arrive. He waves as we pull in. We all walk down to the inlet. It looks very much like yesterday — myriad shorebirds crowded onto the shore by a rising tide — except that today the water is extraordinarily calm. It almost glows, mirroring the brightness of the clouds. We wander along the shore to the spit.

We sit well back from the water at first, enjoying the birds from a distance. Adrian points out the different species for our benefit: western sandpipers, small, with black legs; and least sandpipers, smaller, farther up in the vegetation, with yellow-green legs. A light steady drizzle begins to fall, so gently as to be hardly noticeable, except that our raincoats are soon dripping. The mountains to the north are invisible now, shrouded in cloud and rain.

Adrian wants to get some pictures. He goes down on his belly and squirms forward across the mud. But then he has to back up to get the birds in focus. He's making no attempt at concealment whatever, no blind at all. There are birds all around him, apparently unfazed by this apparition in their midst.

Encouraged by his example, Shirley and I decide that we should get a little bolder too. If he can do it, so can we. Closer to the edge of the water, I lay out my insulated pad for us to share. Then we hunker down, a raincoat bivouac, huddling for warmth.

It's raining steadily now and a breeze from the south ruffles the water. Fortunately it's not cold. It does feel a little odd, sitting calmly in the open, getting rained on. One's natural impulse is to head for cover. (What's the expression? *Too dumb to get out of the rain.*) But it's not uncomfortable once I get used to the idea. In fact, the sight and sound of rain on the water is unexpectedly soothing, almost meditative. It echoes the musical tinkle of the birds talking among themselves. They have calmed down, too. Most are standing quietly, one foot tucked up. Occasionally they rouse to shake the water off their feathers.

I feel increasingly content simply being here, present, not doing anything in particular. In fact, I can't think of a time I've felt more thoroughly engaged with the natural world, part of the whole. The birds seem to think so, too. They had moved away when we took our places, but now they're back. Before long I have western sandpipers foraging within arm's length. A tiny least sandpiper is feeding a couple of centimetres from my right boot. I feel accepted. And so we possess ourselves in patience, waiting for events to come to us.

Eventually the tide begins to draw away. The birds grow restless. Soon it will be time for them to move back onto the newly exposed flats and continue feeding. And the southeast wind is rising, rising.

Suddenly it's over. The moment is past. The flocks are taking wing. Shirley's knees have had enough and I'm aware of the water trickling down my back. I suddenly remember that there are places I have to be, chores that need doing.

Over the water, a great flock of western sandpipers whirls back and forth, like smoke in the wind. The weaving mass flashes white one moment, dark the next, as the birds show first the undersides of their wings, then the upper surfaces, all in perfect unison. Higher and higher they dance, up and down, around and around, finally over the trees and out of sight, lost in the mist.

Western sandpipers belong to a cluster of species, small shorebirds familiarly known to birders as "peeps." Typically found in large mixed flocks, peeps forage across mudflats and sandy beaches, sometimes venturing up into the shoreline vegetation, picking and probing for food with a busy sewing-machine motion. They are all tiny birds, sparrow-sized except for their long narrow wings. Western sandpipers, for example, tip the scales at a scanty 23 to 36 grams (a golf ball weighs about 46 grams).

In winter, western sandpipers occupy an enormous range of coastline on both the Pacific (from British Columbia to Peru) and the Atlantic (from New England to Brazil). In contrast, their summer breeding range is surprisingly restricted: northwest Alaska and the Chukotskiy Peninsula in extreme eastern Siberia.

Perhaps the species was a casualty of rising sea levels at the end of the last ice age. Alaska and the Chukotskiy would have been part of the same land mass during the Pleistocene. Little groups of human nomads making their way across the Bering land bridge during those brief Arctic summers must have flushed great flocks of breeding western sandpipers. Then came warmer temperatures, melting glaciers and the ever-encroaching ocean. Perhaps migrating western sandpipers, even now, are chasing some dim ancestral memory of a great, rich, sea-level plain, long gone beneath the waves — a sort of avian Atlantis.

In springtime, when the world's entire population of western sandpipers gathers into one narrow corridor for the migration north, they become the most numerous shorebird on the Pacific coast of North America. Major estuaries host from a quarter to one million individuals at a time. The birds move from estuary to estuary in stepping-stone fashion: San Francisco and Humboldt bays; Bolinas Lagoon; Gray's Harbor; the Fraser, Stikine, Fox and Copper river deltas; Cook Inlet; Redoubt and Kachemak bays. The males pass through first, hurrying northward to stake out territory. Females follow.

Westerns nest on well-drained tundra from sea-level plains to low mountain slopes. They prefer mixed habitat: ridges and hummocks of heath, with wetlands — marshes, pools and lakes — close at hand. Suitable country can support a high population, with each pair defending a tiny territory, cheek by jowl with their neighbours. They feed mostly on insects, foraging both wet and dry heath. Western sandpipers are monogamous. The female lays four eggs, usually hidden under bushy cover. Both sexes share incubation duties, beginning when the last egg is laid. Twenty-one days later, the eggs hatch.

The chicks are extremely precocious, able to move about and start foraging for themselves almost immediately. Mom abandons the family shortly afterwards. Dad chaperones the kids until they fledge, just nineteen days after hatching; then he takes off too. (Which makes for an interesting reversal of human norms: not only does Mom leave Dad to care for the kids, but when the kids grow up, their parents leave home.) From then on, the young of the year must fend for themselves. They migrate on their own.

Let me repeat that, in case you missed it. The youngsters, barely a month old, make that first migration entirely on their own. The first southbound adults turn up on the beaches of western Vancouver Island in late June. Adult migration peaks in mid-July then falls off rapidly. The first juveniles show up in late July, how they manage is anybody's guess. Research suggests that bonds forged during first migration may last a lifetime, with little flocks from a given year-class feeding together, migrating together, perhaps breeding together.

Most southward-migrating western sandpipers follow a trans-Pacific route that carries them out across the Gulf of Alaska to a landfall somewhere in southern British Columbia or the northwestern United States. That means thousands of kilometres across open water at altitudes up to five and a half kilometres. Pretty much the whole population funnels through the Pacific northwest before dispersing to wintering areas on both sides of the continent.

Now, here's the really interesting bit.

A group of ornithologists around Rob Butler at the Canadian Wildlife Service gathered and analyzed data on weight loss and gain during migration. They considered the length of stay at various rest stops. They calculated the energy costs, hence expected weight loss, for various legs of the journey. And they discovered — even assuming maximum stopovers at estuaries, maximum rates of fat deposition and minimum energy expenditure during flight — that the birds could not possibly cover that vast distance in calm conditions and maintain their body weight. I'll put that another way. Flying in calm conditions, absolutely still air, birds that started the trip in tip-top shape should theoretically lose weight steadily en route and arrive on the breeding grounds fatally emaciated, literally skin and bones. All those birds that can be seen arriving on breeding territory in fine shape would need to deposit fat at impossible rates — three grams a day or more, compared to observed rates of one-third to one gram a day — to make the trip in calm conditions.

It was a mystery.

The sandpipers couldn't possibly be taking on enough calories to cover the rigours of the trip. How were they managing? Well, as any manager knows, when revenue isn't sufficient to sustain the enterprise, the only alternative is to cut costs and improve efficiency. Westerns reduce drag by flying in flock formation — birds in flocks may gain up to five kilometres an hour over birds flying alone. Sandpipers also get more free speed by travelling at high altitudes — flight speeds increase five percent for each additional thousand metres of altitude.

Most significantly, they harness the winds, exploiting the favourable, avoiding the unfavourable. Air masses circulating counterclockwise around low-pressure systems in the Gulf of Alaska are the key to their strategy. In spring, birds can ride southeasterly winds along the eastern edge of a convenient low, up the Pacific coast and into Alaska. After breeding, they hop aboard the same circulation, this time riding it down the west side of the low, going far out over the Pacific and then back toward the west coast of British Columbia. They may even be skilled, like hot-air balloonists, at exploring different altitudes to find the most favourable wind. Truly they are "wind birds." The distances are so immense that they dare not battle a headwind: they don't have sufficient metabolic fuel. When conditions are favourable, they can stay aloft for days at a time, covering thousands of kilometres at a stretch. But if the wind blows contrary, they must get out of the sky and find someplace suitable to rest, stoke up and await the next ride.

The birds are absolutely dependent on those fuelling stations. With their energy budgets so carefully calculated — migration on a shoestring — they cannot afford to lose any sources of revenue. They also need places to come ashore after that long ride over the Pacific, quiet places where they're not going to get chased up and down the beach by someone's dog.

So conserving shorebirds means conserving estuaries. There are, unfortunately, precious few large estuaries on the Pacific coast of North America. And like most forms of wetland, estuaries are being lost to development at a fierce rate, especially in urban areas but also in relatively pristine places like Clayoquot Sound. Even here, development encroaches year by year: more houses, more people, more pets. It's ironic that just when the loss of habitat elsewhere is making the Tofino mudflats more and more important, they themselves should come under pressure.

The front has passed during the night. A pale sun shows through flying cloud-wisps. Along the path, bushes are dripping and puddles have collected in the hollows among the roots. All is quiet, except for the songs of woodland birds.

Again, the tide is perfect, high and full. Water laps calmly on the shore. But nobody's here. The inlet seems vacant, empty, almost desolate.

I walk toward the spit. Nothing rises from the water's edge or from the grass. The rain has erased all tracks from the mud. Except for the odd feather or bit of down clinging to the wrack along the high-tide line, there is no trace of the great flocks. They might almost have been, beginning and end, an illusion.

At the tip of the spit, I put up one solitary bird, a dowitcher. Perhaps there's something wrong with it; couldn't keep up and so was left behind. It flies off, calling, alone and forlorn.

I sit down and scan the inlet with binoculars. Far out across the water toward Browning Passage and Meares Island, a small flock of something semaphores its presence with a brief flash of white. The flock turns and vanishes into its surroundings. Otherwise the place is deserted.

I wait and watch until the tide starts to withdraw. The sun picks up strength. The air is warm. But nothing comes in. The birds are gone.

It's another instance of nature's flywheel moving relentlessly forward. Yesterday's flocks will be far to the north now. Some of the birds that were here two days ago may already have reached the Stikine River delta in Alaska. This is life in the fast lane. Shorebirds have long migrations and short breeding seasons. By mid-summer, less than two months from now, they'll be stopping here again, *on their way back south*. They haven't a moment to lose.

I understand that, but I miss them all the same. The sun is shining; there is promise of a beautiful day ahead. But life seems to have left this place. The excitement has gone elsewhere, following those birds. I imagine the flocks flying northward over a wild landscape of rugged mountains, dark forest, deep fjords. They are heading toward the danger and promise of an Arctic summer. I wish I could go. It's a bad case of *zugunruhe*, migration restlessness, cabin fever. I feel it every springtime and every autumn too.

Perhaps that's why shorebirds have such strong resonance, even for people with no particular interest in birds. They appeal to the nomad in all of us. They awaken some restless vestige of an earlier time — not all that long ago, in the great scheme of things — when human beings had not yet abandoned migration.

Shigi tōku
Kuwa susugu
mizu-no uneri Kana

Afar, shorebirds are flying,
Near, the water ripples,
Washing the hoe

— Buson, 18th century;
from Peter Matthiessen, *The Wind Birds*

June

QAWƆCÅMIL
(Salmonberries Moon)

June is wet and windy. While the rest of the country moves toward summer — T-shirts and shorts and barbecues at the beach — the west coast of the island sometimes seems not merely to have stalled in late spring but to have reverted all the way to winter. Maddening. The sun, when we catch a glimpse of it, rises and sets far to the north. The trees wear full summer foliage. The salmonberries are ripening. And the calendar tells us that summer solstice is upon us; that we stand, in fact, on the threshold of the long march back into winter. All before we've had the least real taste of summer. It's not fair.

In the old times, late spring must have been a relatively quiet season with the people biding their days, waiting for better conditions. The weather was still too stormy to venture much onto open water. If the harvest of herring and roe had been poor, it could be a hungry season. Men spent their days preparing gear and doing ritual for success. A fisherman might go hunting for octopus that he could dry for halibut bait. Women gathered seafood along the shore, shellfish or eelgrass roots, or went into the forest for cedar bark and withes, perhaps the first salmonberries. Hair seal were hunted especially in late spring, though they could be taken at any time. Paddling cautiously, the hunter and his steersman hoped to catch a seal unawares. Feeding undisturbed, the animal would often re-surface quite close to where it had last been seen. When he spied it again, the hunter cast along the surface of the water, snagging the animal in the barbed fork of his double-pointed harpoon. Sea otters and porpoises were also taken; sometimes even sharks. Sea lions were less sought after and fur seals were never hunted before contact with Europeans.

June is still a waiting time, a preparing time. Commercial fishermen, the few that remain, still ready their gear for the summer season ahead. All of the many businesses catering to tourists put the final touch on their operations, anticipating the July rush. The school year is almost over. Children, barely under control, fret in their classrooms and dream of summer days to come.

Siwash Cove
Flores Island

HANGING ON FOR DEAR LIFE: The Intertidal Zone

The south coast of Flores Island is a long series of sandy beaches separated by rocky headlands. At low tide, beaches merge with one another and it's possible to skirt the headlands on packed damp sand. High tide separates the beaches and each headland becomes an obstacle. Most have some sort of trail across them, but each upward step saps energy and each downward step demands extra care, especially in wet weather when the footing is slippery.

I had originally planned to cover the entire distance between Marktosis, on Matilda Inlet, and Siwash Cove in one go. But delay follows delay and by the time I finally step onto the dock at Marktosis, my low tide has long since come and gone. Even so, I remain optimistic. I'm so far behind schedule that high water has also come and gone. The tide is falling again and the days are long in June. I can still cover a good part of the distance to Siwash Cove before nightfall.

It isn't until I step onto the first beach that my confidence begins to ebb. The light on the mountains across the channel is the light of early evening, soft and rosy. The tide is still much higher than I'd hoped. I begin to hurry, just a little, recalculating distances and possibilities as I go.

Then comes the rain.

The first shower catches me before I reach the end of the second beach. I glance up, surprised to find clouds overhead, then hasten to take cover beneath a nearby tree as the drops come thick and fast.

Perhaps it's only a passing shower; the islands in the channel are still bright with sunshine. Valuable minutes tick away. Eventually the rain eases off, sunshine returns and I emerge from shelter ready to tackle the trail across the next headland. Under normal circumstances, a moment's rest in such a lovely spot would be pure pleasure. But I'm impatient, irritated by the delay.

The second shower catches me before I'm halfway down the third beach. Somehow I cannot or will not admit that it's truly raining — the sun shines brightly — and I soldier stubbornly onward. By the time I make shelter at the next headland, I'm drenched. This little downpour also passes, but the light is fading and so am I, feeling the strain of physical effort and cold. My hike is turning into a forced march. Queasy and shivering, I abandon all ambition and begin to consider options for a crash landing.

I know there is sheltered ground at Kutcous Point though I normally avoid it. Kutcous belongs to the Ahousaht people: it's a reserve, private property. They quite properly resent hikers and kayakers who camp there without permission. But maybe this time, given the circumstances, they'll forgive the trespass.

Darkness overtakes me as I set up the tent and I have to finish the job by flashlight. I know it would do me good, but I haven't enough energy to cook a proper supper. I just want to get inside my sleeping bag to warm up. A couple of handfuls of trail mix will have to suffice. As if to mock me, the tide is now falling fast. While I lie shivering, the moon rises near-full and beautiful over the mountains, casting a path of glimmer across the water. So much in life depends on tides and timing.

My timing isn't much better in the morning. The tide is already rising by the time I finish breakfast. Fortunately the Kutcous River is still low enough to wade. I cross just upstream of its mouth, carrying my boots and trousers. A night's rest has restored my optimism. The sun is working hard to burn away the clouds; I entertain high hopes of blue skies to come.

But it is not to be. Little by little, clouds recover their lost ground. Skies are grey and the tide is rising fast when the first rain of the day catches me at the long beach on Cow Bay. I abandon my tattered illusions, call a halt and put on full rain gear: pants, jacket, hat and gloves. It's just as well that I do — high tide forces me repeatedly into the wet vegetation behind the beach. More than once I lose the trail and find myself swimming through salal and salmonberry.

It's late afternoon when I finally stagger out of the forest at Siwash Cove. The sandstone reefs and tide pools I've come so far to see are there at last. But I'm too tired and wet to care. Shelter is the urgent priority now, and food; I'll need something more substantial than trail mix tonight. Also a fire for warmth and cheer. By the time these necessaries are taken care of, darkness and more rain are upon me. Exploration of the intertidal zone will have to wait.

That term "intertidal zone" refers to a narrow strip of coastal ground that is alternately immersed in seawater and exposed to air, according to the rise and fall of the tide. The amount of exposure varies, depending on where in the intertidal you happen to be. Some areas are immersed in water or, at the other extreme, exposed to air for just moments at a time, once or twice a year during the very highest and lowest tides. Others are covered and uncovered twice a day. It is a difficult environment for living things, being neither marine nor terrestrial, sea nor land, but something in between.

The intertidal zone is never very wide but it is enormously long, running for thousands of kilometres along all the world's complex and varied coastlines. The zone comprises a great variety of more or less distinct sub-environments. There are high-energy coasts, exposed to the pounding of ocean surf. There are low-energy, sheltered coasts. There are rocky shores, sandy beaches, cobble beaches. There are estuarine environments in sheltered situations where fresh and salt water mix. The west side of Vancouver Island boasts a full range of different environments supporting some of the most fertile intertidal communities in the world.

Every environment is fascinating in its own way, but for me — and for most visitors to this part of the Pacific coast — a visit to the high-energy rocky intertidal, with its tide pools and surge channels, is the very acme of natural-history experience. Tide pools are natural aquaria: potholes and crevasses excavated from the rocky shore by surf and populated by a mosaic of living plants and animals, even a few sub-tidal organisms, raised up into the world and conveniently displayed to save us landlubbers the trouble, danger and expense of scuba diving.

The best tide pools are found on the most exposed sections of coast, which stands to reason. Those are the very places where time and heavy surf have had the best opportunity to excavate the necessary cavities. The natural turbulence also assures a constant flow of nutrient-rich, oxygen-rich water and a heavy growth of interesting sea life. The southwest corner of Flores Island is the epitome of such places. Siwash Point separates Siwash Cove and Cow Bay from the island's highly exposed west coast: a spectacular landscape of ocean swell breaking across broad sandstone shelves and the reefs beyond. It's a fantastic place for tide pools, for intertidal life in all its forms. And the low tides of solstice will show it to best advantage.

The weather is only marginally brighter next morning, but at least it's not raining while I make breakfast and plan my campaign. Low tide is scheduled for 9:30 in the morning, less than an hour away. I'm here, I'm ready; my timing seems to be coming around at last. I pull on my damp rain gear and go down to see what I can see.

Perhaps because they are so extensive and well defined, I have a stronger sense than usual of descending through the various levels of the intertidal: first the relatively barren rocks high on the shore, then a zone of fucus or rockweed, then beds of California mussel with all their commensals, then a band of brown kelp and laminaria, and finally, at the very lowest levels, a zone of red kelp. With every step downward, the living communities become more prolific and complex. The very lowest levels are a riot of colour and form, life upon life.

The relative harshness of physical conditions along the upper edge of the intertidal zone — including long exposure to drying air, sun and terrestrial predators — ensures that those species tough enough to handle the conditions have no shortage of elbow room. Closer to the low-tide line, where physical conditions are less demanding, the number and variety of potential tenants rises sharply.

At the very lowest levels, competition for space is extreme. Any exposed surface is rapidly colonized. Aggressive species overgrow or evict their milder-mannered competitors. It's very common to see plants and animals, "epibiotics," staking out living space on the surfaces of other plants and animals, down to the most minute. A large mussel — which is, itself, attached to other mussels — is host to a colony of limpets and barnacles. Looking closely at the limpets, you find that their shells support colonies of even tinier barnacles. And those tiny barnacles might show a growth of bryozoans or sponges, which in turn support a growth of algae or bacteria. And so it goes. *Big bugs have little bugs upon their backs to bite 'em. The little bugs have smaller bugs — and so ad infinitum.*

Tide pools are a separate element in this scheme. All plants and animals in the intertidal zone are marine organisms — they require salt water for essential biological functions such as respiration and reproduction. Many are fully alive only when immersed, but all are capable of surviving some exposure to air. Some, like the tiny snails and barnacles inhabiting the barren rocks of the upper intertidal, can get by with just the occasional splash at high tide. Others, like the various sea stars, are less intrepid and can survive only briefly out of water.

By capturing and holding the retreating seawater, tide pools offer a sanctuary for plants and animals that cannot tolerate long exposure to air: sea stars, urchins, nudibranchs, fish and suchlike. Some of these are sub-tidal organisms that cannot survive any time out of water at all. In tide pools we get a glimpse of the world beneath the waves.

At the very lowest limits of the intertidal, especially at a very low tide, I often have the eerie feeling of walking — trespassing — on the temporarily exposed floor of the ocean. A parting of the waters. Plants and animals glisten wet. Great ocean swells break across the point of the reef. The water sucks and gurgles, impatient to rise and reclaim its own.

But in that stolen moment, you can glimpse a host of wonders. Here are sea urchins, purple and red, big as cantaloupes and armed with long spines. And sea stars: sun stars, leather stars, bat stars, blood stars, the usual ochre sea stars and a sunflower star half a metre across. Anemones: strawberry anemones, enormous green Pacific anemones, aggregating anemones called *elegantissima* in Latin for their multi-coloured tentacles. Seaweed of all sorts, slick and iridescent, brown and purple, smooth or covered in warts, stalked and plumose; one species looking for all the world like miniature palm trees. Here, truth is stranger then fiction, the fantastic is normal.

Ocean waters along the west coast of Vancouver Island are among the world's most fertile. The riot of life on this reef has built up to take advantage of those rich resources. But it's a harsh and challenging environment. The rise and fall of the tide, the pound and shear of the surf, the intense competition for space and food make life difficult. This tension between rich resources and difficult conditions has driven and shaped the evolution of whole communities of bizarre, seemingly experimental life forms. Here — as in deserts, the high Arctic and at mid-ocean vents — the harshest environments seem to support the most remarkable ecosystems.

I ambush a large urchin, snatching it from a tide pool before it has time to panic and grip the rock. Urchins are related to sea stars, sand dollars and sea cucumbers, all members of the order Echinodermata. All echinoderms have spines — "echinoderm" means "spiny skin" — and tube-feet with little suction cups to grip substrate or prey. In urchins the spines are highly developed; in sea stars the tube-feet are more prominent.

I balance the dripping creature atop my open hands like some pagan offering and wait for developments. There is a certain urgency to air exposure for both urchins and sea stars. All echinoderms operate their tube-feet with a water hydraulic system. High and dry, they leak hydraulic fluid profusely. Before long they are helpless. At first I feel only an agitated, random movement of the creature's sharp spines on my bare skin, as if it were paralyzed by indecision, not knowing which way to turn. (Being a radially symmetrical animal, of course, it faces a greater variety of choices than you or I would.)

Then a thought comes to me. I tilt my palms slightly, with the idea of giving the urchin something to think about, a basis for informed decision-making. Sure enough, a moment later the movement of spines becomes coordinated and the creature begins to stilt-walk downhill off my hands, presumably with some dim idea of regaining its tide pool. I save it the fall and put it back where I found it.

Here also are some of the largest green Pacific anemones I've ever seen, thirty centimetres or more across. Anemones are the quintessential intertidal animal. Exposed by the falling tide, they fold in upon themselves to conserve moisture, becoming featureless green pillows. But underwater they bloom, appearing very much like flowers at first glance. Their bright colours, their shape — a cylindrical stalk surmounted by a disc fringed with fleshy, petal-like tentacles — even their name suggests some sort of plant.

But no, they are animals — predators, in fact. The fleshy tentacles are armed with tiny stinging cells, nematocysts. Any item of prey, say a small crab unlucky enough to find itself dumped on the anemone by a wave, would first be captured and immobilized by the tentacles, then conveyed to the mouth, a navel-like opening in the middle of the upper disk. The anemone is a simple animal. Its digestive system is a sac, a cavity in the stalk, with just the one opening. A few hours after a meal, any indigestible bits are regurgitated through that same opening.

I dip my hand into one of the pools to touch a large specimen. It feels soft, smooth, almost gelatinous. When I brush one of the tentacles, it clings firmly to my finger. Happily, my skin is too thick to be penetrated by the nematocysts and I am safe from any toxins. But it has a firm grip on me and, as I watch, the other tentacles bend slowly inward to lend assistance. The entire disc starts to tilt my way and the creature's mouth begins to gape. I think, *this beast is planning some outrage.* Enough is enough. I gently disengage myself from its tentacles.

Normally when we think of predators, we imagine active, speedy animals pursuing nimble prey. But the anemone is a predator that cannot hunt, a predator fixed in one spot. Of course, it doesn't need to pursue its prey; the surge and flow of the ocean delivers tasty items right into its waiting arms. What the anemone really needs is some means to avoid being snatched from its sheltered tide pool by that same surge and battered against the rocks or cast onto the shore, where it would surely die.

Here is a fantastic reversal of the normal situation, like science fiction. Imagine a world where the wind is so strong that lions are rooted to the ground, waiting for zebras to come whirling by on the breeze. And yet the anemone's situation is absolutely typical of the intertidal zone, where every living thing is hanging on for dear life. Sea stars and urchins grip the rock with their tube-feet. Barnacles are cemented in place with a miracle mortar of their own manufacturing. Snails and limpets adhere by the suction of broad muscular feet. Mussels anchor themselves with byssal threads. A wicked-looking kelp crab, spider-like and big as my hand, grips the seaweed with its hooked feet — while looking me straight in the eyes and daring me to try something funny.

Incidentally, anemones — like many intertidal animals — are surprisingly long-lived. These larger individuals may be a hundred years old. I'm looking at old-growth tide pools.

There could be no tide pools, of course, without the rise and fall of the tide. There could be no intertidal zone and, without the intertidal, perhaps, no life on dry land. The intertidal must have been a cradle for terrestrial life. One can see the process at work even now, as periwinkles and shore crabs evolve toward an independence from mother ocean.

If Earth were alone in the universe we would have no tides. But we are not alone. For one thing, we have a sister planet: the Moon. Earth and Moon orbit one another, ponderously, every 27.13 days. We like to say that the Moon orbits the Earth, but this is pure chauvinism. Both bodies are in motion, revolving around their common centre of mass.

Now, here's the essential fact. The direct gravitational attraction of the Moon varies, depending on proximity. The closer you get to her, the stronger the attraction — as with any good relationship. But the distance between Earth and Moon, orbiting one another, is determined by the average gravitational attraction between the two entire masses, the centripetal force. Only at the very centre of Earth will the actual gravitational attraction of the Moon be equal to the centripetal force in both strength and direction.

That is why any particle — a molecule of seawater, say — on the side of the Earth closest to the Moon, feeling the disproportionately strong gravitational attraction of the Moon relative to centripetal force, will have a slight tendency to draw away from the planet, falling toward a tighter orbit around the Moon. At the same time, any particle on the opposite side of Earth, away from the Moon, feeling an urge for independence under the relatively feeble gravitational attraction, will have a tendency to drift outward into a looser orbit.

These effects are not confined to molecules of seawater. The whole planet creaks and flexes like a wooden boat in a storm. We ourselves are less firmly attached to the surface when situated on the near or far side of Earth, relative to the Moon. In fact, when the Moon is directly overhead, we are lighter by about ten milligrams — weight watchers take note — feeling that tendency to drift up into a tighter orbit.

This disparity between the actual gravitational attraction of the Moon and centripetal force is why Earth's oceans tend to gather slightly on the side facing toward the Moon and simultaneously on the side facing away from it. As Earth spins on its axis, one revolution every 24 hours, those gathered waters move like two great waves across opposite sides of the slow-rolling surface of the globe.

On Chesterman Beach, an observer experiences the crest of each wave as a high tide. The trough between two waves is low tide. The west coast of Vancouver Island sees two complete tidal cycles every 24 hours and 50 minutes: the time required for Earth to complete one revolution plus a bit extra to account for the Moon's progress along its orbit. In a little less than 25 hours, our observer would see the tide rise, fall, rise and fall again. Each point in the cycle would come about 50 minutes later than it did the previous day — just as the Moon crosses the zenith about 50 minutes later every day. The two tidal cycles in a 25-hour period are often unequal. One cycle will last much longer than the other and be more extreme, rising to a higher high, falling to a lower low.

Over the course of a lunar month, there is a pronounced cycle of variation in the height of the tides coinciding with the phases of the Moon. Both full Moon and new Moon bring extreme high and low tides, known as spring tides (which has nothing to do with the season; they occur all year-round). Quarter Moons are associated with much more moderate tides, called neap tides.

These variations reflect the degree of coincidence between the gravitational influences of Moon and Sun. The Sun is orders of magnitude more massive than the Moon. But because it is so much farther away, it has much less influence on tides. Picture a ball ten metres in diameter. A blue ball, sitting on the sand. If that were the Earth, the Moon would be a ball, white, 2.5 metres in diameter, about three hundred metres down the beach. The Sun would be an enormous blazing ball a full kilometre in diameter. But it would be about 115 kilometres off the coast; and gravitational attraction declines as a function of distance squared.

We don't even see distinct solar tides. What we do see is the influence of solar gravity on lunar tides. At the quarter Moon, when Sun and Moon lie at 90 degrees to one another relative to Earth, their gravitational effects tend to cancel each other. Tides are moderate. When Sun and Moon are aligned on the same or opposite sides of Earth, at new Moon and full Moon, gravitational effects coincide and extreme tides are the result. The most extreme tides of the year occur near summer and winter solstices, when the greatest declination of Sun and Moon, relative to Earth's equator, coincide.

There are myriad other factors. Salinity (extra salty water is heavier), water temperature (colder water is heavier), the natural resonance of enclosed bays, the elliptical shape of the Earth–Moon orbit, the wobble of the Earth on its axis, all influence the tides. Even weather plays a role. A strong ridge of high pressure or a wind blowing from land to sea squeezes water away from the shore, forcing a low tide even lower. A low-pressure air mass or a wind blowing toward shore will pile the high tide higher. A combination of low pressure and strong onshore winds during a severe winter storm can create an extreme high tide called a storm surge.

Time presses. The water is rising. Reluctantly I leave the reef and head for the broad sandstone shelf on the outside of the point. It stretches out of sight up the coast, inviting exploration. If only I had more time and better weather. These shelves are like petrified beaches, relatively fragile, now cracked and worn by the surf. There are potholes and cavities everywhere, an embarrassment of riches for the naturalist.

Along its outside edge, the shelf ends abruptly in deeper water. Swells heave suddenly upward and break across the ledge. I look up to see a gray whale surface forty or fifty metres offshore, beside a gleaming brown mass of bull kelp. The animal blows, sounds, then appears no more.

This must be a terrifying place in a winter storm, utterly exposed to the open ocean, deafening with the roar of crashing waters. And yet the shelf is covered by layers of life, acres of surf grass swirling beneath the waters like long lengths of women's hair, a gathering of mermaids.

Down where the rising tide has not yet covered these beds of surf grass, I part the living hair and discover whole communities of tiny plants and animals taking shelter beneath it. The grass must serve like the chafing fibres on a fisherman's net to absorb the wearing, tearing force of the surf, sparing the living community underneath from fierce pounding.

This too is typical of the intertidal. While they hang on for dear life, the inhabitants must also find some way to parry or avoid the pounding and tearing of the surf. Many — like the little creatures beneath the surf grass — simply hide. Anemones find deep pools and crevasses. Purple urchins create little foxholes in the soft rock, rasping away with their spines, generation after generation. Other organisms, like sea palms and gooseneck barnacles, rely on tough fibre and flexibility. Still others, like acorn barnacles, put their faith in homes and armour plating. Chitons and limpets go for smooth and elegant streamlining.

A masterful few combine their strategies. Mussels are a prime example. Each mussel is held in its place by a strong network of byssal threads, anchor cables sufficiently elastic to give a little beneath the pounding of the waves. Each shell is beautifully streamlined to cut the water. And that elegant shell is also tremendously strong, to resist the pounding of the surf.

I love to watch the waves roll in upon the shore, one after another, endlessly. There are always waves on the windward coast; you can count on it. From the gentle lapping of summertime seas to the thunderous rip and roar of winter storm surf, the ocean is endlessly restless. Night and day, the waves roll in, year in and year out.

Waves are born of the wind. Perhaps a bit of turbulence or friction at the water's surface raises the first tiny wavelets, something the moving air can get a purchase on. The wind — bouncing off the crest of each ripple and creating a pocket of low pressure beyond — both pushes and pulls the growing waves along.

It's important to understand that a wave is not an object but an event: a progressive or propagating impulse, like a line of dominoes falling one after another. It is the event that moves down the line, not the dominoes. Water moves only in a limited way with the passage of a wave. A chip of wood floating in deep water traces a vertical circle as each wave passes: it moves toward an oncoming wave, rises to meet the crest, is carried on the crest back over the top of the circle, and ultimately subsides toward its original position as the wave passes on.

A wave's size depends on three factors: wind speed, wind duration and wind fetch, the unobstructed distance over which the wind blows. Winds of similar force can create very different seas, depending on how long they blow and whether they blow across open ocean or sheltered inside waters.

Once waves are properly established, their momentum can carry them forward even when the wind stops blowing. The smallest of these propagating waves, the ones that aren't damped out by inertia and friction, tend to be absorbed into larger, faster-moving waves. Big waves can outrun the storm that spawned them. That is why the surf comes up on Chesterman Beach hours before the arrival of an approaching low-pressure system. These big propagating waves, ocean swell, can travel vast distances with minimal loss of energy. Oceanographers have tracked ocean swells with wavelengths up to two hundred metres from their natal storms in Antarctica across the entire Pacific Ocean to journey's end on the coast of Alaska.

Such long-distance travel has important implications. The Pacific Ocean is so enormous — the largest single physical feature on the planet, twelve thousand kilometres across, a third of the Earth's surface

— that there are always storms brewing somewhere in the vastness. Each storm stirs up a crop of big waves. Because ocean swells can travel such tremendous distances, the west coast of Vancouver Island is within range of waves from storms all over the Pacific — winter storms in Antarctica, a typhoon south of Japan, low-pressure systems in the Gulf of Alaska. No wonder the waves never cease on the outer coast.

Eventually, rising water drives me from the shelf. I splash my way to shore across a living carpet. And this is a troublesome thing, the dark side of any intertidal excursion. I am thinking now of the lives that my little visit has disrupted or ended. I'm not so much worried about the urchin, the anemone or the other individuals I've inconvenienced with my peeking and prodding and plucking. They'll be okay. What concerns me is the inadvertent mayhem: the physical impact of simply being here and walking around in this delicate place. Every step I've taken has been a catastrophe to some of the unlucky plants and animals beneath my feet. The intertidal is thick with life. In places, the rocks are completely covered with layers of living organisms. It's impossible to avoid stepping on something.

That may not be so grievous as it sounds. Intertidal plants and animals can take a good deal of abuse. They have evolved, after all, to survive the pounding of winter surf; I'm pretty mild punishment by comparison. But I'd be kidding myself to pretend that I haven't damaged anything here this morning. I console myself that at least I'm no real danger to the ecosystem. There is a high natural rate of turnover here; it's part of the way this community functions. My contribution to murder and mayhem is trivial. Each species is more than sufficiently fertile to make up the losses.

So perhaps I can persuade myself that my impact has been minimal. But what about the next visitor? And the next? At what point does it become too much? When does a piece of countryside begin to die the death of a thousand cuts? How do we know when to stop before we love the place to death?

I have no answers, so I go on, stepping lightly as I can. But it's a concern. Part of me feels an anxious urge to make the world aware of all this beauty, to advertise the special places. I want masses of people to see these wonders for themselves, fall in love with the place and then speak up on its behalf when the time comes. But another part of me hopes urgently that the world stays far away from fragile treasures like Siwash Cove.

With only a few minutes of useful low tide left, I move quickly back across the point to explore the sandy beach inside the cove, a very different environment from the sandstone ledges or the reef. Here there is nothing firm for living things to hang onto. The sand is in constant motion. Not much life shows on the surface; most of the inhabitants seek protection and security by retiring deep into the beach. And yet, in its own way, this environment is almost as fertile as the rocky intertidal.

I stop to investigate a tiny mound, volcano-like, marking the exhaust end of a ghost shrimp burrow. The ghost shrimp makes his living by excavating a U-shaped tunnel in the sand, setting up a current and filtering out any edible particles that come floating through. I start digging. A few moments later, elbow-deep in muck, I come up with the occupant, pale and pink, one claw enormously enlarged, looking more shield than weapon. When I drop him back into the wet sand he is gone almost instantly, burrowing out of sight in fear of his life.

It's dog-eat-dog in the intertidal zone. If a seagull catches that little shrimp exposed, it will be the end of him. So just as they must hang on for dear life and protect themselves from the pounding of the surf, every living thing in the intertidal must also have one or more strategies to avoid becoming a meal for somebody else. They armour themselves with shells, disguise themselves with camouflage, hide deep in rock, make themselves unappetizing with pincers and poison.

Another excavation unearths a polychaete worm, large and writhing, like a small blind dragon. Its fanged proboscis flicks the air with switchblade suddenness, seeking meat. I dip into the beach once more and come up with a bent-nosed clam, tight shut, as self-contained and immobile as a stone.

With the tide rising faster now, I move along to a small area of cobble beach — coarse stones instead of sand — on the east side of the cove. It's just the sort of place you might find cockles, clam-like bivalves, edible and choice. *Singing cockles and mussels, a-live a-live oh.* I've left it a little late, but a quick search through the rocks turns up one small specimen at the very edge of the rising water. Not a keeper.

As I prospect, I become aware that there are shore crabs every-where among these rocks, hundreds of them. Wherever I turn, I catch the flicker of furtive movement from the corner of my eye. I hear the dry whisper of many claws scratching on stone. Whenever I overturn a rock, the space below is alive with dozens of little crabs scrambling for shelter. I wonder, how do they all get enough to eat? It's a rich environment, but the competition for food must be awesome. Every creature has some trick, trade or tool to help it capture a share of the goodies. The cockle, a filter feeder, has nets of mucous. That ghost shrimp has its burrow, a plankton trap. Crabs have pliers for hands and mouths full of little jointed knives. The beach is an enormous living garburator. Anything the least bit edible falling among these rocks is quickly dismembered and gnawed into nothingness by the whispering hordes, invertebrate piranhas. And anything too small to interest the crabs will feed the cockles and barnacles and ghost shrimp. I resolve to stay on my feet and keep moving.

As I cross the cove, heading back to camp, I notice an odd, soup-bowl-sized mound in the sand. It turns out to be the day's most spec-tacular find and a fitting finale to my explorations: a lovely moon snail, spectacular creature, with a white shell as big as my fist and a great round foot the size of a dinner plate. I've surprised it at lunch, the enormous foot embracing a doomed clam. Typical. It's dog-eat-dog in the intertidal.

By late afternoon the weather is showing sure signs of improvement. The sun even ventures out for a time, allowing me to shed some rain gear. I take a book out onto the warm rocks. Later that evening, I even get my hoped-for glimpse of the Moon rising over Cow Bay.

Next morning is cloudy but dry. I eat breakfast and have the tent down in good time. By nine o'clock, I'm saying goodbye to Siwash Cove. My timing is good. The tide is still falling. Having made all the mistakes on the way out, I'm sure of my route now; the return trip is a piece of cake. Before long the sun is shining in earnest. Halfway to Cow Creek, an eagle launches from a small spruce and sails across the beach in front of me, white head and tail gleaming in the sunlight. I pack my rain gear away.

The tide is very low. I manage to avoid most of the headlands and cross the creeks where they flow shallow across the sand. I make it back to the Kutcous River in three hours and don't even have to take my boots off to wade across. Another hour sees me back in Marktosis, with just enough time for a cold soft drink from the store before I catch a water taxi back to Tofino.

One and a half days going out. Four and a half hours coming back. So much in life depends on tides and timing.

July

TA'ATOKÁMIL
(Drifting Moon)

Summer comes in July. The weather settles down at last and we're treated to a week, or two or three, of sunny weather before the fog starts rolling in. It's the only real summer we get. By mid-month the shorebirds are passing through in large numbers, southbound now, already making the return trip from their breeding grounds. Some of those little fluff-balls foraging along the shore have just completed a non-stop crossing of the Gulf of Alaska. Already the berries are ripening, storing up sugar while the sunshine lasts.

In the old times, summer was the season to take advantage of calm weather and bountiful harvests: sea mammals, berries, fish. Halibut were fished on the great banks far offshore, sometimes so far from land that the coastline was hidden below the horizon and the fishermen navigated by aligning the peaks of high mountains. They started from the beach at night to take advantage of early morning calm, paddling in the dark, five, ten, fifteen, twenty kilometres offshore. They fished the morning hours and headed home at midday, their canoes often loaded to the gunwales with enormous fish, some almost as big as the men themselves. Many other foods were gathered in summer. Salmonberry shoots appeared in late spring, followed shortly by the salmonberries themselves. Then came thimbleberries, huckleberries, blackberries and salalberries. Most were eaten fresh, but mashed salalberries could be formed into cakes and dried. Sea urchins were collected most often in the summer although they could be had at any time of year.

In modern times, too, July is the season to take advantage of the calm weather and bountiful harvests of summer. With school out, visitors throng to the outer coast. Business booms for the hospitality and recreation industries. Art galleries and restaurants thrive. Everyone is busy, busy, busy. Even so, July is a time for hanging out on the beach, getting onto the water, maybe barbecuing a salmon in the warm light of evening, sweet cedar smoke drifting on the breeze, full moon rising.

Clayoquot Canyon

THROUGH A GLASS DARKLY:
A (Very) Brief Voyage of Oceanic Discovery

> *I must go down to the seas again, to the lonely sea*
>   *and the sky,*
> *And all I ask is a tall ship and a star to steer her by;*
> *And the wheel's kick and the wind's song and the white*
>   *sail's shaking,*
> *And a grey mist on the sea's face, and a grey dawn*
>   *breaking.*
>
> — John Masefield, *Sea Fever*

Time for mid-morning coffee break. Our boat, an eight-metre Boston whaler christened *Eco*, gently rises and falls in a low glassy swell from out of the northwest. The twin Evinrudes at the stern, 120 horsepower apiece, are quiet, thank goodness, at least for the moment. The sun is shining. The air is warm. It's a lovely day, all in all, but I'm nervous as a cat.

Some 40 nautical miles (75 kilometres) to the east, a jagged line of peaks edges the sky: the mountains of Vancouver Island. In one sweeping glance I can compass all the country between the Brooks Peninsula to the north and Cape Flattery in the south, a spectacular view. At the same time, the panorama is oddly unfamiliar and disorienting. Only the tallest peaks, the highest mountains are visible. The lowlands are gone, sunk, drowned in the ocean's swelling horizon. I would find it difficult, without resorting to the *Eco*'s charts, to place Tofino or Clayoquot Sound with any accuracy.

Also drowned, shrouded in everlasting darkness, is the ocean floor far below — 1,000 metres straight down, a full kilometre. And according to the chart, the bottom is still sloping downwards, steadily downwards, away from the coast. Ten nautical miles (18.5 kilometres) west of our present position, the depth is 1,500 metres; 20 nautical miles (37 kilometres) to the west, it reaches 2,000 metres. And 45 nautical miles (83 kilometres) west of our present position, a total of 80 nautical miles (148 kilometres) off the coast of Vancouver Island, the depth reaches 2,500 metres, two and a half kilometres. That's where I run out of chart.

The idea of our little open boat, alone on the vastness of the ocean, suspended high above the abyss; the thought of that terrible gulf of water below; the thought of breaking through the surface, drifting into the darkness, down and down, the intense cold, the terrible pressure, the scavenging mouths — it has all given me an acute case of the heebie-jeebies. I can feel the skin creeping between my shoulder blades. To add to my paranoia, a slight breeze has just come up, breaking the glassy surface of the sea into little ripples.

Already this month's expedition, though brief, has been an education. For one thing, I have learned the true meaning of the word "landlubber." I yearn toward those mountains the way a child in darkness yearns for the light, the way a man who has strayed unexpectedly into a dangerous foreign territory yearns for the comfort and safety of home, even as he wonders if he'll live to see it again. You get the picture. I'm not entirely comfortable out here. I have an inkling of how the first astronauts must have felt looking back on Earth, safe and familiar, from the perilous remoteness of the Moon.

My fears — though perhaps a tad exaggerated — are not altogether unwarranted. This *is* an alien environment and we are here purely on sufferance. Everything looks safe enough now, but the slightest change in the weather could put us in deadly peril. We have ventured so far from land this morning that a sudden blue-sky gale could easily stir up dangerous seas in the time we'll need to regain sheltered waters.

So what brings us out here, so far from land, tempting fate? Officially, the *Eco* and her crew are engaged on a survey of pelagic birds and mammals for the Strawberry Isle Research Society (SIRS). I'm beginning to think of it as extreme bird-watching; if we make an error, we lose our lives.

My own reasons are more complicated. There is an element of whistling past the graveyard. Perhaps I have exaggerated my fears, but I am genuinely nervous about deep water: an ancient and atavistic anxiety that I wish to confront. And this trip is my opportunity to cultivate a better acquaintance with that most alien and powerful of all west coast environments, the open ocean.

The marine influence is everywhere in Clayoquot Sound. From broad sandy beaches still wet with salt water to remote mountain peaks buried beneath immense depths of snow, ecosystems are shaped by their proximity to the Pacific Ocean. To understand the land, we must understand the ocean.

But my attempts to understand the ocean remind me of an old folk tale. A group of blind men describe an elephant. The first, feeling the animal's broad flank, declares: *An elephant is like a wall.* The second, with his hand on one of the tusks, replies: *No, an elephant is hard and round and sharp, like a spear. Nonsense,* says a third, taking up the trunk, *an elephant is like a snake.* And so it goes.

When it comes to the Pacific Ocean, I am like those blind men. I speak from such a limited perspective: *The Pacific Ocean? Yes, of course. It's an undulating, mirrored, liquid surface under the sky, sometimes smooth as molten metal, sometimes fierce with great crested waves.*

I'm not alone in my blindness. As far as our species is concerned, the ocean is still a great unknown, unknowable. We're stuck on the outside trying to catch a glimpse of whatever's happening inside, the events and the circumstances, perceived as through a glass darkly. A thorough, systematic exploration is still beyond us. Even in an age of scuba, deep-diving submersibles and remotely operated vehicles, we are like blind men reaching into the darkness: touching here, touching there. It is said that, until very recently, we knew more about the surface of the Moon than we knew about the seabed that covers three-quarters of our planet. Research in oceanography and marine biology has burgeoned in recent decades, but our knowledge is still fragmentary. The deep trenches and the vast abyssal plains might as well be on the far side of the Moon.

Perhaps that's why the ocean sciences, even now, have such a hypothetical and speculative flavour to them. Ideas are based on a handful of actual observations and an imperfect understanding of even the most general processes at work. That air of uncertainty, of mystery, of the vast unknown, lends oceanography, marine biology and all related sciences an aura of real excitement. These are frontier enterprises. Fundamental discoveries are yet to be made; anything is possible. Even ordinary people can make a contribution simply by being present and keeping their eyes open.

Consider our skipper, Rod Palm. Palm first came to Tofino in the late 1960s as a commercial diver. Besides the bread-and-butter work of his trade — laying cables, servicing vessels, salvaging wrecks — he was responsible for many of the pioneering underwater archaeological surveys, chiefly historical shipwrecks, along the west coast of Vancouver Island.

In 1985 he began to photograph killer whales as a hobby for the late Dr. Michael Bigg of the Pacific Biological Station in Nanaimo. It was soon apparent that killer whales, also known as orcas, were spending more time in Clayoquot Sound than anyone had suspected. It also happened that most of the killer whales Palm photographed were of the little-known transient variety.

Transient orcas are even more mysterious than resident orcas. Small groups of transients forage unpredictably over a wide range along the Pacific coast from California to Alaska, following no discernible pattern. They show up here and there, then disappear for months at a time. All this is in contrast to the larger resident pods that spend their summers, year after year, feeding on salmon in relatively limited and predictable ranges through sheltered waterways along the east coast of Vancouver Island and Puget Sound. The transients feed almost entirely on marine mammals. The two types never mix, even when they meet. They look different, even to human observers. They sound different. They even differ genetically.

By 1991, Palm's hobby had escalated into a full-blown project. The Strawberry Isle Research Society (SIRS) was established to support the work on killer whales, funded in large measure by Tofino's whale-watching businesses. More than ten years later, the society continues to generate important data on transient orcas and has undertaken a variety of other natural history projects: studying the use of Grice Bay mudflats by gray whales, monitoring sea lion populations and so on.

In October 1993 the society ran its first monthly pelagic survey. The surveys were launched to gather information on birds living on the open ocean off the west coast. It's a more or less completely different set of birds from those commonly seen inshore. But the abundance of marine mammals encountered on the first few outings suggested a wider scope for the project.

Surveys employ small fast boats to run a transect beginning at Wilf Rock near the south end of Vargas Island and extending 34.5 nautical miles (64 kilometres) directly offshore, traversing the continental shelf and terminating above the abyssal plain. Each trip follows a standard protocol. There is a port-side spotter and a starboard-side spotter, who is also the recorder. Birds and mammals sighted along the way are recorded, numbers and species, along with the distance from shore, water depth and temperature. All observations are collected on audiotape for later transcription.

It's all straightforward enough in summer, when the weather is calm. But in the early years, Palm and his observers did winter trips as well, making their dash to the abyss in the lulls between storms. Considering my jitters on a sunny day in July, I can hardly bear to think what some of those trips must have been like. They were surely risking their lives, but nobody had taken a serious look at pelagic activity off the west coast during the winter. It was virgin territory, a chance to make a contribution, for anyone with the courage to go out and take a peek.

Better them than me. I'd be the klutz who got washed overboard.

The four of us gather bright and early at the marina: Palm, myself, one volunteer from the research society and another from the Canadian Parks Service (the surveys are funded in part by Parks Canada). The park warden and the society volunteer, both practised observers, will be the trip's official spotters. Palm will coordinate the effort and pilot the boat. I'm just along for the ride, an extra pair of eyes.

While introductions are being made, our skipper fusses with the boat, making certain that everything is shipshape. He checks the safety gear and tops up the fuel tanks; big outboards have an inordinate thirst at open throttle. Satisfied, he cranks the engines and backs *Eco* from her slip, the two Evinrudes rumbling and burbling happily at the stern.

The idea is to avoid tempting fate any more than we have to. We are not blindly trusting to luck, we're entertaining a calculated risk. Reports on weather and ocean conditions have been carefully checked and considered. Our trip is timed to catch the early calm of the day. And we have the benefit of Palm's long experience.

The town always seems subtly different when seen from the water. There is an odd feeling of disconnect; even the sound of traffic comes as if from a distance, echoing. We are barely away from the dock and the *Eco* has already become part of something else entirely, even though life ashore goes on as usual. We've passed through the looking glass. No wonder it always seems a moment of truth, no going back, when a big ship casts off and draws away from the quay. At that instant, the ship and every soul aboard leave the familiar domestic world behind and enter into something much less familiar, and rather less predictable.

The harbour has its own wide-awake life first thing on a bright summer morning: commercial fishing boats heading for the fuel docks; water taxis; smaller boats of all descriptions; water people commuting from the islands to their town jobs, or maybe just looking for a morning cup of coffee, a bit of gossip; kayakers off on a day trip; traffic to and from the fish plants; floatplanes landing and taking off on early flights for Marktosis or Hot Springs. It's a different world, the water world, and water people don't always see eye to eye with their staid counterparts up the hill, settled like barnacles in rigid immovable structures.

Palm follows the usual route out Duffin Passage, around Felice Island, westward toward Moser Point on Vargas Island. From there we turn into Father Charles Channel and make for Wilf Rock. As the *Eco* roars down the channel, our view of the wide horizon opens up, ocean and blue sky, everything a little hazy in the humidity. I sense the land falling back, falling away, releasing us into the ocean. Ahead, a vast emptiness, a new and alien environment. Wilf Rock looms abruptly to the right, the edge of the known universe. As we speed past, our skipper enters the course on his Global Positioning System and starts the clock. We're underway.

The continental shelf off Vancouver Island is a broad triangle, 40 to 50 nautical miles (75 to 90 kilometres) wide in the south, narrowing to almost nothing at the north end of the island. The seafloor is a simple extension of the terrestrial landscape along the outer coast, gradually sloping westward, a rolling topography with low hills and hollows to break up the monotony.

At a depth of about 90 fathoms, 160 metres, the bottom begins to fall away more steeply. This is the shelf break. Beyond it is the continental slope, rough country, sculpted into troughs and ridges that run more or less parallel to the break. The whole drowned landscape is divided by numerous canyons running like river valleys toward the abyssal plain. Our destination this morning, the Clayoquot Canyon, commences in a nearly vertical 30-metre head wall about 35 nautical miles (65 kilometres) southwest of Tofino. From the base of that precipice, the canyon drops steeply over the first seven or eight nautical miles (13 to 15 kilometres) of its run. At a depth of about 800 metres, the slope eases off and the gulch covers another two to three nautical miles (3.5 to 5.5 kilometres) before opening onto a broad submarine terrace at a depth of about a thousand metres. For all intents and purposes, that is where North America ends and the Juan de Fuca plate begins.

All that submarine real estate is buried, frosted, by a thick layer of guck — "sediment" is too dignified a word — a combination of inorganic and organic particles. The inorganic particles are bits of pulverized rock: sand and gravel nearer shore, silt and clay farther out. The organic particles are bits of dead plants and animals. A third type of sediment, less common, forms when various chemicals — carbonates, phosphorites, manganese and the like — precipitate out of super-saturated solution around hydrothermal vents or submarine volcanoes. In the strictest sense, the bottom of the ocean *is* guck. In some places — not here, this coast is too young — sediments are thousands of metres deep, accumulated over millions of years.

On the continental shelf, different substrates — rock, cobbles, sand, and mud — foster different biological communities. Rocky substrates tend to attract more species of fishes and invertebrates, perhaps because of the greater variety of habitats. Fewer species make their homes in sand or mud.

Bottom-dwelling organisms, benthic species, can be divided into two categories: the epifauna living on the seafloor — crabs, brittle stars and bottom-dwelling fish — and the infauna that live below the surface — clams or burrowing worms. Benthic communities change with increasing depth and changes in the sediment. In very shallow waters with a sandy bottom, amphipods, nudibranchs and gastropods predominate. In deeper water, shrimp and urchins are more abundant. On the mud bottoms of the middle and outer continental shelf, deposit-feeding polychaete worms prevail.

Food is scarce down on the abyssal plain. Population densities and growth rates are low, compared to the continental shelf. But the diversity of benthic species is relatively high, perhaps because food items cover such a wide range: everything from the minute organic particles that rain down out of surface waters to the mighty sunken carcass of a whale.

Pelagic organisms — life forms living free of the bottom, suspended in the water column — also fall into two categories: plankton and nekton. Plankton are plants (phytoplankton) and animals (zooplankton) that swim poorly or not at all. Individually minute for the most part, planktonic organisms occur in enormous numbers, drifting helplessly with the current. Phytoplankton are the ocean's primary producers, forming the base of the food chain and producing 95 percent of the ocean's oxygen. Zooplankton feed on phytoplankton, providing a vital link with the rest of the marine food web.

By contrast, nektonic organisms are active swimmers: marine reptiles, pelagic birds, marine mammals and pelagic fish. The last group in particular tend to be fast, streamlined predators: herring, anchovies, hake, salmon, sharks, mackerel, tuna.

> *I must go down to the seas again, for the call of the*
> *running tide*
> *Is a wild call and a clear call that may not be denied;*
> *And all I ask is a windy day with the white clouds flying,*
> *And the flung spray and the blown spume, and the*
> *sea-gulls crying.*

By all accounts — Palm and his two observers, that is; I'm new to the game — the ocean is extraordinarily calm. A real piece of luck. The surface is glassy, unruffled by the slightest breeze. As always, there's a bit of a swell running, but very moderate, low and rounded. At first it seems of little account. I begin to relax. But as we continue, minute after minute, roaring along in our monotonous, straight-line flight from the land, the swoop, dip and pound of the boat breasting across the swell becomes tedious, then irksome, then positively wearing. I begin to yearn for something, anything, that might give us an excuse to stop for a moment. But despite the excellent visibility, there isn't much to see, and what we do see is pretty ordinary. Gulls. A harbour porpoise. Some common murres. A rhinoceros auklet. Marbled murrelets. And, as common sense might dictate, the farther we travel from shore, the less we see.

Meanwhile, Palm counts off time and distance. At about 25 nautical miles (45 kilometres) out, just as we approach the edge of the continental shelf, we begin to notice a change in our immediate environment. The water is clearer, more blue than green. Suddenly we're seeing many more birds, and new species, too. Sooty and Buller's shearwaters — visitors from nesting colonies in New Zealand. Pink-footed shearwaters from Chile. A northern fulmar, perhaps already back from breeding grounds in Alaska. Sabine's gulls. Fork-tailed and Leach's storm petrels. A cluster of different alcid species in good numbers: rhinoceros auklets, Cassin's auklets, tufted puffins. An enormous black-footed albatross cruising by on wings spanning two metres.

There is also a whole batch of new and interesting creatures in the water. In short order we spot a blue shark, then a mola-mola or sunfish and, moments later, a fur seal sleeping on the surface, rocking on the bosom of mother ocean. We are almost on him before he wakes with a start and is gone.

I remember reading, years ago, a set of early science fiction stories, written when human beings were still absolutely bound to the surface of the planet. The great gulf of air above was mysterious to a degree we can hardly imagine nowadays. This particular set of stories speculated that the first aeronauts would encounter a whole new world of creatures in the upper atmosphere, great diaphanous things living their whole lives beyond human view high above the Earth. Such ideas seem fantastical and quaint to us now, but this morning I feel as if I have experienced something similar.

Instead of progressing through an ever more impoverished emptiness, we have come upon a thriving community, flourishing unsuspected — at least to me — way out in the middle of nowhere, hidden below the dip of the horizon. I know about pelagic birds, of course, but I am genuinely unprepared for the richness of this ecosystem and the abrupt transition: all this busy activity after kilometres of nothing.

There are some good and logical reasons why certain spots — and not others — in all the seemingly featureless monotony of the ocean should enjoy such a profusion of life.

Remember that seawater absorbs light very effectively. Red and violet wavelengths vanish from the spectrum within a few metres of the surface. Blue-green light penetrates somewhat farther, but even then, in the clearest of ocean waters, less than one percent of ambient light reaches 100 metres down. Along the west coast of Vancouver Island, water is even more opaque. Sunlight penetrates only 15 to 35 metres, depending on angle of incidence, cloud cover, the smoothness of the surface, the amount of plankton in the water. Closer to shore, where plankton grows more luxuriantly and runoff from the land may cloud the water, penetration is reduced even further. Most of the water in the ocean — and most of the seafloor beneath it — is utterly dark.

And where there is no light, there is no life; at least, no plant life. The great majority of marine organisms live within a few metres of the surface, in the euphotic zone where light is sufficient to support photosynthesis. At greater depths, living biomass declines sharply. Plant growth is restricted by the lack of light, and animals must depend on organic matter sinking out of the surface layers.

(Hydrothermal vent communities, which harvest the chemical energy in geothermal water, are a notable exception.)

Even where light is plentiful, a lack of nutrients can limit plant growth. Nutrients tend to precipitate out of the euphotic zone. In the tropics and subtropics especially, surface waters can become so nutrient-impoverished they have been called biological deserts.

So here's the problem in a nutshell: Surface waters have light but lack nutrients. Benthic environments have nutrients but lack light. What's needed is a magic re-combination, a bringing together of ample sunlight and nutrient-rich water.

Circulation is the key. Oceans are not static. Seawater circulates through the abyss much as air circulates through the atmosphere. In fact, atmosphere and ocean are closely influenced by one another. Surface currents in the open ocean are driven mostly by atmospheric circulation. Surface waters to depths of between 100 and 500 metres are subject to mixing by wind and wave. They are also much influenced by the sun's warmth and by freshwater runoff. Not surprisingly, surface water temperatures and salinities vary widely from place to place and season to season.

Surface waters and deep waters tend to be separated by zones of relatively abrupt transition: the thermocline (a boundary marking a change in temperature), the halocline (change in salinity), the pycnocline (change in density). Deep waters are colder, saltier and denser than surface waters.

In deep waters, circulation is driven by differences in density between different masses of water, based on variations in temperature and salinity. This thermohaline circulation is more or less analogous to density-driven circulation of the atmosphere. The density of seawater, like the density of air, varies with temperature: the cooler the water, the denser and heavier it is. The density of seawater also varies with salinity: the more salt dissolved in a volume of water, the denser and heavier that water is.

In the Arctic and the Antarctic, bitter cold air chills the ocean's surface. Cold water sinks to the bottom and oozes south or north. In tropical seas, evaporation often exceeds precipitation, especially where there is relatively little freshwater input, as in the Mediterranean and the Red Sea. That extra-salty water, like the extra-cold water of higher latitudes, sinks to the bottom and oozes outward.

Oceanographers envision water masses of different density flowing over one another, very much like air masses in the atmosphere. Cold Arctic water moving south across the floor of the Atlantic Ocean is bulldozed upwards by even colder (–1°C) Antarctic water heading in the opposite direction. Great lenses of salty water spill from the Straits of Gibraltar and slip through the darkness toward America, spinning ponderously like enormous flying saucers.

Meanwhile, surface waters may be moving in entirely different directions. The shear between these water masses may engender deep turbulence and eddies analogous to storms in the atmosphere — midnight hurricanes in the depths of Hades.

This thermohaline circulation has been likened to a conveyor belt that circulates water in a haphazard way among all the world's oceans. Eventually deep water returns to the surface — cold water mixes with warmer water, salty water mixes with fresh — and is then carried by surface circulation back toward the poles or into high-evaporation tropical seas, to begin the process again.

This resurrection of nutrient-rich deep water into the euphotic layer is the magic I spoke of; this is the key to the profusion of life we're seeing here this morning.

In the winter, when winds blow predominantly from the south, all surface waters off the west coast of Vancouver Island drift northward. In summer, the prevailing wind is from the northwest. Under the influence of freshwater runoff, surface waters close to shore persist in drifting northward. But surface waters along the outer part of the continental shelf reverse direction, moving southward with the wind. When the wind blows strong and steady, a thin layer of surface water, perhaps twenty metres deep, tends to drift away from the coast toward the open ocean. And as that surface layer slides away from shore, it is replaced by deeper water welling up from below.

That cold, nutrient-rich water, exposed to sunshine, supports a vigorous growth of phytoplankton and, by extension, a vigorous growth of all other species up through the trophic pyramid: zooplankton, fishes, marine mammals, birds. Even human beings. It's quiet here this morning. But if we'd come at another time, we might have found the ocean busy with a whole commerce beyond the ken of landsmen: trollers out for salmon, trawlers fishing for hake, the processing ships of several nations.

The break is over. Palm takes his temperature reading and water samples. Then he fires up the Evinrudes and we head for shore, to my very great relief. Swoop, dive and pound. Swoop, dive and pound. For a long time, the view changes little. It's a comfort, nevertheless, to be heading toward a visible destination rather than the empty horizon. Slowly, ever so slowly, the lower mountains — Catface, Lone Cone and the others — rise into view. Gradually I'm able to orient myself again.

This must be what it feels like to return to Earth from space. First, the whole planet spread out before you on a vast scale, an unlimited range of choices. Then, a gradual zeroing in toward more and more specific destinations: decisions made, options forgone. Finally, touchdown, contact, a rolling stop and there you are, fixed in the landscape once more, all your choices behind you.

As we race toward shore, the wind builds steadily, feeding my anxieties. *Nothing to worry about*, I tell myself, *just a little daytime heating*. But before long, the sea has grown so choppy that Palm is forced to throttle back to save wear and tear on the boat. *No sweat*, I think, *we're almost there. Almost there*. I can even see the lowlands now: Estevan Point and the coastal plain behind Long Beach.

Bird-watching has become pretty hopeless. In fact, we almost miss the gray whales. "There!" shouts one of the observers. Then I see it: a fading plume of vapour hanging above the dark blue of the roughened sea. Our skipper hauls the *Eco* around and we move in for a closer look. A cow and calf with another adult, keeping them company, perhaps. Palm reports the sighting by radio. Before long, a hard-hull Zodiac roars out to join us, scattering waves left and right. Both boats hang back a respectful distance, just watching, as the whales plough stolidly northward. Floater-suited tourists, bright orange and yellow, snap pictures that will show little more than a vast field of rolling waves. We leave them to it and head in.

I can see coastal landmarks now, natural and man-made: Wilf Rock, Echachis Island, Lennard Island light station. Even the big resorts. Finally I can spot individual houses. Then we're flying up Father Charles Channel past Vargas Island, around Felice Island, across the harbour and into the slip at the marina. The warden jumps out to make the lines fast. I force myself to slow for a dignified exit. I express my gratitude to Rod for allowing me to come along, for giving me a glimpse of something totally new and unexpected. I say goodbye to my fellow rovers. And then I take that tremendous first step from the boat to the dock. Not terra firma exactly, but close enough for me.

> I must go down to the seas again, to the vagrant
>   gypsy life,
> To the gull's way and the whale's way, where the wind's
>   like a whetted knife;
> And all I ask is a merry yarn from a laughing fellow-rover,
> And quiet sleep and a sweet dream when the long
>   trick's over.

August

SATSÅMIL

(Chinook Salmon Moon)

Deborah at the bake shop calls it Fogust. Too true.

While the rest of the country is enjoying the brilliant blue skies and hot days of late summer, the weather here is often more reminiscent of November: dense fog on the beach and a cool breeze blowing off the ocean. Not every day is foggy, of course, but even when the sun does shine that ocean breeze keeps temperatures down. It's like living in front of an open refrigerator. Bikinis are out; wetsuits are in. Fortunately the fog soon dissipates as it drifts over the warm land; often when the beach is foggy, the sheltered side of the peninsula and the inlet are blessed with sunshine.

Perhaps that explains why, in the old days, late summer was the time when people moved from their summer villages on the open coast to autumn villages far up the inlets where salmon would be harvested in the fall. Such moves were major undertakings. Belongings were stowed in baskets and boxes, roof and siding planks taken from summer houses and lashed across two or more canoes to form a raft, everything piled aboard for a trip that might involve a day or two of paddling. Imagine a whole household on the move, all ranks and ages: white-haired elders, middle-aged men and women, youths, children, babies lashed into their cradles. All the work notwithstanding, it must have been a happy, festive time, a welcome change of scene with lots of laughter, talk and singing. Upon arrival, the planks were unlashed from the canoes and set onto house frames already in place from previous seasons; the boxes and belongings moved in; mats laid out; the household quickly re-established in its new location. Already the first chinook salmon, huge silvery fish, were gathering off the mouths of their natal streams: the quietness of sheltered waters broken from time to time by the explosive flash and splash of jumping tyee.

In modern times, fog or no fog, visitors keep coming; maybe some of those people even travel here to get away from the heat. August is another busy month. Nevertheless, there is a subtle change in the air, a sense of revolving seasons. Perhaps it's the fog and cool temperatures, or the alders, which have already started to shed their leaves, but while the rest of the country is still deep in summer, the west coast of the island seems already to be looking toward autumn. But don't worry; summer is keeping back a little something for later.

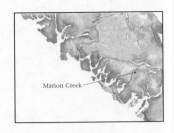

Marion Creek

IN THE SHADOW OF THE PACIFIC:
The Mountains of Clayoquot Sound

Who goes to the Hills goes to his mother.
— Rudyard Kipling, *Kim*

The Marion Creek road was constructed in sections over a period of about ten years, beginning in the early 1980s. Carved from mountain rock, it represents a huge investment of money and effort, all simply to reach the big timber up the valley. Logging equipment rolled in. Trucks rolled away with load after load stacked high — western redcedar, western hemlock, Douglas fir, white pine — until much of the land on either side of the creek had been stripped of trees. And then the whole show was abandoned, a disposable road, just another write-off.

What's left of the scenery along the creek is unlovely. The ground is littered with slash and broken wood, half rotten now and silvered with years of hard weather. The clearcuts here are finally "greening-up," as industry representatives like to say: from a distance the valley floor no longer has the raw, red appearance of freshly logged country. But up close, the new growth turns out to be a tangled jungle of shrubbery: salmonberry, thimbleberry, young alders and three or four species of *Vaccinia* (huckleberries and blueberries). The bright purple of blooming fireweed is everywhere, and already downy seeds ride warm thermals into the sky. The first few conifer saplings, mostly western hemlock and western redcedar, are coming up through the shrubbery, but a man's lifetime will have to pass before this piece of ground supports anything close to proper forest again.

I wish it were not so. I wish the road had never been built. I wish the big trees along the creek were still thriving, the valley green and lovely. But I must admit — and there's irony aplenty in this — that if it weren't for the Marion Creek road, I probably wouldn't be here today. This is my highway to the high country.

Clayoquot Sound boasts extensive areas of alpine terrain. That comes as a surprise to most visitors; the sound is much better known for its beaches and low-elevation rainforest. But if you look at any springtime aerial photo of western Vancouver Island, you'll notice the expanses of white contrasting so vividly with the surrounding green. That's late-lying snow, lots of it, precisely delineating subalpine and alpine. Together they comprise a considerable proportion of the landscape.

But access isn't easy. The bush is thick and the terrain precipitous. Extended cross-country hikes are impossible. For most people, residents and visitors alike, the mountains remain distant abstractions, beautiful but untouchable, forever out of reach — kind of like a girl I knew once.

Unless you have a road.

The goal of today's outing is a cluster of alpine meadows at the head of Marion Creek. I'm hoping to reach the summit of a modest little peak that happens to be the easternmost corner of all Clayoquot Sound's watersheds. With luck it should afford me a grandstand view of the whole show.

I might also catch a glimpse of the Vancouver Island marmot, one of the world's rarest mammals, with fewer than a hundred surviving individuals. Most of the population lives on a single alpine ridge 70 kilometres away to the southeast. The mountain where the species was first recorded is even closer, just 47 kilometres due east. It's entirely possible that they're present on the ridges above Marion Creek; the habitat is perfect.

And there should be wildflowers aplenty. In the mountains of western Canada, early to mid-August is prime time for alpine flowers.

I'm expecting a good show. All in all, I anticipate a most enjoyable day. But I'm also conscious of a severe conflict of interest. Here am I, relying on this road, built with the sole purpose of hacking up the countryside, to seek an experience of natural wilderness that would otherwise be beyond my reach. Makes me uneasy.

The route isn't what it used to be. When the supply of timber ran out and the heavy equipment departed, engineers took steps to deactivate the right-of-way. To check erosion they placed ditches *across* the steeper grades. Taking it slow, I make it over the first couple of ditches, both relatively shallow. But the next set are deeper. The little truck bottoms out once, twice, three times before I decide to stop pushing my luck. The nearest tow truck is sixty or seventy kilometres away. I'll walk the last stretch to the pass.

I make good progress following what's left of the road, first along one side of the valley, then across the creek and up the other side. Eventually the grade crests; I have achieved the pass. Behind me, Marion Creek drains north into the Kennedy River and ultimately into Clayoquot Sound. Ahead, a different watershed altogether: the Effingham River flowing south into Barkley Sound. The pass lies at the end of a deep narrow valley with spectacular mountain scenery all around. To my right, the jagged peaks of the Mackenzie Range; to my left, a forested ridge that hides today's objective. The whole landscape has been glacier sculpted. It doesn't take much imagination to see ice oozing over the ridge above me, breaking into a chaos of jagged sky-blue crevasses and cascading in ultra-slow motion down the slope.

I need to leave the road here and find a route across the clearcut, then up the steep forested slope of the valley wall. The lowest alpine meadows are already in view, seemingly not too far up the mountain. But the heavy fabric of forest on that slope almost certainly conceals some near-vertical terrain. Even from down here I can see rock showing through the greenery. I can't begin to plot a route. I'll just have to start somewhere and hope for the best. But where? On the far side of the clearcut, a small creek comes down through the forest: as likely a point of departure as any.

Just getting across the clearcut turns out to be a modest ordeal. The thick brush is an impediment and the piles of rotting wood give way underfoot, suddenly and at the most awkward moments. I'm mortally afraid of catching my leg in a deep hole: it would be easy to sprain something, perhaps even break a bone. It's hot and dry out here, gravel pit conditions, which is ironic considering the site of Canada's wettest day is less than twenty kilometres away and the site of Canada's wettest year is even closer, about twelve kilometres as the crow flies. If there were a weather station in this valley, it would doubtless be measuring similarly tremendous winter precipitation. But with the forest cover gone, rock and gravel quickly give up moisture to the summer sun.

By the time I reach the forest I'm breathing hard and drenched with sweat. The mosquitoes love it, as does a swarm of blackflies up from the creek. Under the trees, the difference in climate is astonishing. A few steps beyond the harsh light of the clearcut, the air is cool, the moss underfoot a delicate green, the earth still moist. At this altitude on the very eastern edge of Clayoquot Sound, the undergrowth is relatively thin, mostly well spaced *Vaccinia*. The forest is also relatively open, the trees well-spaced: western hemlock and a few scattered amabilis fir, both familiar from lower elevations. Already the western redcedar is giving way to yellow cedar, a graceful tree with pale, yellowish, aromatic wood, almost reminiscent of juniper.

Every tree in sight shows evidence of snow-creep. Trunks come out of the ground at a distinct downhill angle before curving upright into the normal posture. In winter, the snow on this slope slips gradually downhill. Young saplings are forced downhill, too, like reeds in a heavy current, until they grow tall and sturdy enough to resist the pressure and resume their upright growth.

Cooler now and rested, I start to feel more hopeful about this mad expedition. I forge upward. My optimism is short-lived; soon I'm climbing on all fours, only to discover that the undergrowth is affected by snow-creep as well. Every shrub grows downhill at an angle calculated to resist my progress. The stems, growing prostrate along the ground, are slippery little railings. My boots get no purchase, I slither and slide.

A rock face bars the way. I bear right, looking for a route, then left, almost to the creek. I'm feeling more and more anxious. If the creek bed were any steeper, it would be a waterfall. At this time of year, the water-worn rock is coated with a slippery summer growth of algae. It would be death, pure and simple, to set foot out there. I'm just thinking about turning back when I discover an apparent break in the cliff above. It's close to the creek, but not too close; steep, but not too steep. I move upward to check it out.

An hour later and I'm still climbing. It's almost too good to be true. What are the odds that there would be a clear route through this obstacle course of broken rock? And what are the odds that I would find it so directly?

It would be easy to dismiss the whole experience as a happy fluke, but it's far from the first time I've run into this sort of thing on my wilderness travels: a route opens up where there seemed none; obstacles melt away.

Once, paddling back from Hot Springs Cove, a clear moonlit night, I encountered a sudden fog bank. A few strokes carried me into deep darkness, silence, cold. What to do? My destination was kilometres away. I'd never make it in the fog. Neither could I think of any other stopping places nearby. In my memory, the countryside was rugged, dense forest rising straight from the water. Hardly room to stand, never mind pitch a tent. I was tired, wet and absolutely unable to continue. Despairing, I turned and paddled for shore, or so I hoped. At length the fog thinned and a small rocky beach came into view, directly ahead. Across the beach, straight in front of the bow, a clear path led into the forest. Twenty or thirty paces in was a made campsite: a fire ring, a clear space for my tent, a place to hang my gear. It would have been only slightly more eerie to find a fire crackling and supper waiting. I almost wondered if the fog had been conjured up on purpose, somebody wanting company for the night. It seemed just too much of a coincidence to be coincidental.

I could give other examples. Point is, now and then I have the distinct feeling that somebody or something, seeing my need, is laying out solutions where I can find them. Either that or I'm able to plug in, unawares, to some source of knowledge about the landscape, something outside myself. Question is, had I hiked a different direction up this slope, would this path still have opened up to receive me? If I had pointed the kayak in a different direction on that dark night, would a campsite still have been waiting for me? Odd thoughts for a summer afternoon. *There are more things in heaven and earth, Horatio, Than are dreamt of in your philosophy.* (*Hamlet*, Act I, scene V)

More likely, I suppose, my subconscious has been busily sorting out the route ahead, taking in the physical data and analyzing it, while my conscious mind idled along, enjoying the scenery. A perfectly logical explanation. And this is the glacier's old track, wider than the present creek but following the same path, smoothing a way through the rock faces, indifferent to my need. A happy coincidence.

Perhaps.

Western hemlock gives way to mountain hemlock as I climb. The two species share the same soft, lacy foliage, but mountain hemlock needles radiate in all directions from the twig, little starbursts of green rather than the two flattened ranks of western hemlock needles. A lovely tree, characteristic of Vancouver Island's subalpine forests, where life is dominated by a deep, persistent snowpack.

Conditions in subalpine forests — we could think of them as snow forests — are more challenging than conditions in lower-elevation rainforests. Summers are short and cool. Precipitation is very heavy and falls mostly as snow. Winter snowpack accumulates early and stays late, often into summer. But for plants and animals with the right qualities and equipment, subalpine forest is not so bad. Deep, persistent snow may be a problem for smaller plants with their leaves and flowers just a few inches off the ground — false lily-of-the-valley, say, so successful at lower elevations. But trees can laugh at heavy snowfall, provided they can shed the excess accumulations. A tree trunk is an extraordinary adaptation for deep snow, a mast or tower that lifts the plant's foliage high above the deepest drifts. With that problem solved, snow becomes a distinct asset, effectively insulating the ground. No matter what the air temperature, the soil rarely freezes. Roots are safe and trees have access to liquid water through the entire winter. If the weather is fine, evergreens may even be able to photosynthesize a little.

Deep snow is also a problem for some animals. Deer have trouble getting around: their narrow hooves sink in deep and they must retreat to lower elevations. But for others, deep snow is a blessing. Voles forage under the snow by burrowing. The snowpack not only insulates them from the cold, it hides them from predators.

I reach the crest of the ridge, at last, and the slope eases off, thank goodness. The mountainside above the ridge is sculpted into a classic cirque; according to my map, a little lake lies cupped in the hollow somewhere not too far ahead. Wonderful view of the mountains on the far side of Marion Creek. I also glimpse, for the first time, the summit above the cirque. My spirits soar and that old familiar euphoria — call it *rapture of the steep* — comes flooding back. I do love the mountaintop view. No wonder the ancient Greeks placed their gods on Olympus. In the mountains, even mortals enjoy delusions of grandeur.

A few steps farther and I come to the first meadow, outlier of alpine tundra above.

Some ecologists consider this landscape of alpine meadows separated by islands and hedges of subalpine forest to be a distinct type of ecosystem: subalpine parkland. Others see parkland as a simple zone of transition, an "ecotone," between subalpine forest below and alpine tundra above: a mosaic comprising fragments of two different ecosystems. In the hollows, where cold air collects and snow endures into the summer, or on exposed faces, where winter conditions are extreme, trees cannot grow and alpine vegetation takes over. But where conditions are less extreme, trees do grow and fragments of subalpine forest establish themselves.

Meadow vegetation here is dominated by mountain heather, crowberry, partridgefoot and kinnikinnick. Together they represent a distinct botanical community, alpine heath, typical of mild, moist alpine areas. False azalea and dwarf willow have replaced *Vaccinia* as the dominant shrubs. The surrounding evergreens, dwarfish compared to their conspecifics in the forest down the slope, are mostly mountain hemlock and yellow cedar, with a few scattered subalpine firs.

Just as I'd hoped, the mountain is bright with a profusion of wildflowers, particularly where the ground is moist. There is the orange of Indian paintbrush and the purple-blue of lupine; white and pink mountain heathers; yellow arnica and groundsel; white springbeauty and Sitka valerian; yellow buttercup and cinquefoil; the pink and yellow of subalpine daisy and leafy aster; the great green leaves and pale flowers of Indian hellebore and cow parsnip. In the spring there must have been yellow glacier lilies; the seedpods are everywhere.

Alpine areas in Clayoquot Sound bear an uncanny resemblance to those in the Coast, Selkirk and Monashee mountains of mainland British Columbia — all relatively mild, high-precipitation uplands. Though separated by several hundred kilometres of lower-elevation country, they have many plant and animal species in common. I always feel a little uneasy up here. There are no grizzly bears, I know that, but it sure *looks* like grizzly country. I can hardly control the urge to make lots of noise, a habit learned on the mainland where grizzlies still roam and surprises are something to avoid.

As I work my way toward the lake, the meadows open up. The remaining trees, mostly mountain hemlock and subalpine fir, clump together in little islands of dwarf forest. Within each island, a thick growth of branches, tangled and impenetrable, hugs the ground.

Only a few sparse upright stems reach bravely for the sky. Oddly enough, the steep south-facing slope to my left still supports a luxuriant growth of forest. There are full-sized trees growing far up the mountainside.

The idea of tree line or timberline has a cultural resonance and many romantic associations. *Out beyond timberline. The never-summer land.* But there is really no such line. Physical conditions determine the living community. A south-facing piece of ground, sheltered from the wind, well watered but also well drained so the soil can warm quickly in the spring, will grow trees even well up the mountain.

On the other hand, a north-facing pocket of waterlogged soil, exposed to chilly winds, covered with snow and frozen for a good part of the year, may only be able to support alpine vegetation even if it lies far down the mountain. Of course, climbing up the mountain you ultimately reach a point where there are no more trees; and coming down, there are ultimately no more alpine meadows. But it's hardly a line.

By the time I reach the lake, I'm ready for a break. These little cirque lakes are created by the rasping, plucking action of the glacial ice flowing off the surrounding slopes. The ice would have churned about in the basin here, like a stream at the bottom of a waterfall, then oozed away down the mountain carrying the excavated material. In the life of the glacier, this would have been the first place that ice accumulated and the last place it lingered. This is where erosion went on longest.

It's not a large lake, a couple of hectares, but deep and clear. The bottom goes down and down into the gloom. The water is cold and covered in ice for most of the year. There is little plant growth and I see no fish. Maybe a few larval insects; otherwise it seems virtually sterile.

Such stillness seems strange, disquieting, accustomed as I am to the ocean's fecundity and tumult. A quiet, empty lake in the midst of all this Wagnerian mountain scenery. *There be monsters here.* But the water is cool and lovely on a hot day. I risk a drink, and delicious it is.

I have no idea whether there might be *Giardia* or other parasitic nasties in the water, but if you can't drink the water this far off the beaten track, where can you drink? And what kind of world will it be when we can't drink the mountain waters anywhere at all?

The bugs are fierce. Clouds of mosquitoes halo my head. Fortunately the breeze is rising. I find the airiest spot I can and sit down to eat a sandwich. While I eat, I watch and listen for marmots, getting out my field glasses and scanning the rocky slopes behind the lake.

There are six different species of marmot in North America, including the well-known woodchuck or groundhog of the eastern United States and Canada. (*How much wood could a woodchuck chuck, if a woodchuck could chuck wood?*) Vancouver Island is home to only one species, and it occurs nowhere else. Vancouver Island marmots are handsome animals, dark chocolate brown with white muzzles and white patches on forehead, breast and belly. They are a fair size, as rodents go, about the same weight as a big house cat, but pudgier in build. (Well, yes, I've seen cats like that too, but you know what I mean.)

Marmots excavate elaborate burrows and spend much time underground, even in summer when they are mainly active outside in the morning and evening. They hibernate between September and April. Social animals, they cohabit quite peacefully, engage in mutual grooming, get along well with each other. It's quite charming to watch. Sometimes they can be spotted basking on sun-warmed rocks. But the real clue to their presence is a conspicuous contact call: a shrill, far-carrying whistle.

The species is said to prefer open subalpine areas, especially south- to west-facing meadows at elevations above a thousand metres. This little cirque fits the description perfectly. But if there are marmots here, they are keeping quiet and out of sight.

Mosquitoes and blackflies do not encourage relaxation and I have ground to cover. I finish my snack, slip the pack on and move out. I decide to seek a route up through the forest on the left-hand side of the cirque. If I can make it to the crest of that slope, I should be able to work my way toward the summit across the open ground I expect to find up there.

It turns out that the forest above is divided by a network of alpine meadows, invisible from below, each meadow conveniently leading to others. I make good progress, though the steeply sloping meadows have their own hazards. The heather grows thickly and is surprisingly slippery. More than once my feet go suddenly out from under me. I pick my way carefully and wonder what it will be like coming down.

At the top of the crest, conditions change abruptly. Hard to imagine this place in winter when it's exposed to the full blast of hurricane winds coming off the Pacific. But there are a few hints. The only soil, meagre spoonfuls, lies in sheltered crevices between boulders. In places, frost has heaved regular patterns on the naked ground, polygons of smallish stones.

But wherever there is soil and water, plants flourish — a thick summer growth of sedges, lupine, Indian hellebore. Drier ground supports heather and other heaths. The bare rock itself is home to lichens, slow-growing and unbelievably tenacious. Lichens can endure freezing and drying, may take a thousand years to cover a patch less than a centimetre in diameter. But ironically, they are extremely susceptible to disturbance. The same could be said of alpine communities in general. Where plants struggle to survive and grow so slowly, the ground cover is easily damaged and terribly slow to heal.

There is snow up here still, hardened and sculpted by the summer sun, lying in great drifts at the bottom of deeper gullies along the north side of the crest. It is now mid-August. The first fresh snow of winter will fall in September or October. Will these drifts be gone by then? Or are they semi-permanent — nascent glaciers — waiting patiently, dreaming of ice ages past and ice ages to come? At the very edge of the snow, growing from the gravel, are vivid clusters of red-purple monkey flowers. No trees. And no marmots, either, though I pause to look and listen.

Alpine conditions are otherworldly: high winds, heavy snow, intense and prolonged cold. Frost occurs at any time of year. Soils are generally frozen in winter. In places, ground frost, permafrost, can persist year-round. Where the ground is permanently frozen or drainage otherwise interrupted, soils become waterlogged. Most precipitation falls as snow, lying in great drifts or blowing like sand in a desert storm. Growing seasons are extremely short. And yet, in summer, daytime air temperatures can be high. High-altitude radiation is intense. The air is thin and thirsty, soils are rocky; drought is a real problem. Both plants and animals require extraordinary adaptations for survival.

Alpine plants typically assume compact, ground-hugging forms — cushions, mats, rosettes — keeping themselves out of the wind in summer and safely beneath the snow in winter. In fact, most alpine plants keep the better part of their biomass below ground in large root systems, safely out of harm's way. No energy is wasted on excess growth. And almost all alpine plants are perennials. There just isn't time to complete a full reproductive cycle in a single season.

Alpine growth is exceedingly slow, a marathoner's pace, matched to the availability of resources. Endurance is what counts. A plant will not flower unless it achieves carbohydrate levels beyond what is required for simple survival. If weather permits seed production, ripening may not occur until the following summer. If seeds germinate, growth may be so slow that leaves are not produced until the second year. Slow and steady wins the high-country race.

On the other hand, alpine plant metabolisms are supercharged to handle short growing seasons. Even colour counts. Pale flowers, white or yellow, shaped like tiny parabolic reflectors, warm the reproductive organs; insects attracted by the warmth seek out and linger in flowers. Alpine plants can carry on photosynthesis and reach maximum performance at lower temperatures than plants from milder climates. This may be why alpine species do poorly at lower elevations. They run too hot, as a mechanic would say, in warmer environments.

Water loss is a tremendous problem for alpine plants. In summer, rocky soils are low in moisture. In winter, water is frozen and unavailable. The problem is compounded by fierce winds. Hugging the ground and hiding beneath the snow helps plants to avoid desiccation. So do the various compact growth forms. And alpine plants also affect hairy or waxy coverings, further protection from the drying wind.

Where barest survival is so difficult, reproduction may present a well-nigh insurmountable challenge. Most alpine plants are pollinated by insects: hummingbirds are not common above timberline. And since bees prefer warmer temperatures, many alpine plants cater specifically to flies, which remain active at cooler temperatures and lower light levels. In more severe environments, self-pollination is the rule. And where the growing season is just too short for any sort of sexual reproduction, vegetative reproduction with bulbils and rhizomes dominates.

There are no trees in the alpine. They are too profligate.

That marvellous trunk, the secret of success, a competitive advantage at lower elevations, becomes a liability high on the mountain. Wood laid down in trunk and branch represents a major investment of energy. But wood is unproductive tissue; it doesn't photosynthesize. In that sense, trees are inefficient. They have a poor ratio of productive to unproductive tissue. At lower elevations, where energy is plentiful and competition intense, trunks are a good investment. They help the tree to compete. But at higher elevations, where energy is in short supply, they are a fatal extravagance, and no longer adapted to the surroundings: in an environment where sensible plants hug the ground, trees are still foolishly reaching for the sky, exposing themselves to the abrasion of ice crystals and the moisture-sucking winter wind.

Cold isn't an issue. Properly hardened trees can stand very low temperatures. But they need time and warmth to ripen, time to complete their growth, time to lignify cell walls, time to lay down thick cuticles on their needles, time to concentrate their juices so as to resist freezing. And time is terribly short in the alpine.

High on the mountain, trees try to compensate by assuming dwarf forms: krummholz, crooked wood, elfinwood. Trees as thick as my wrist may be hundreds of years old. With no time to produce seeds, krummholz spreads by layering: wherever branches touch the ground, they take root. Higher on the mountain, most of a tree's energy goes into the thick tangle of growth near the ground. Upright stems become more and more slender. At the absolute limit of growth, trees may abandon upright stems altogether, growing prostrate behind whatever shelter they can find.

Animals must also be extraordinarily well adapted to survive and prosper in the high mountains.

Consider the missing marmots of Marion Creek, a wonderful example of alpine adaptation. Marmots are actually ice-age mammals, contemporary with woolly mammoths, short-faced bears and dire wolves, animals that would have been at home in the great polar deserts of fifteen thousand years ago, species that evolved in tundra conditions. It's sobering to think that the Vancouver Island marmot may be headed in the same direction as the others, toward extinction. I might as well be looking for woolly mammoths on this mountain.

Marmots are essentially gigantic squirrels, a clear example of Bergmann's Rule: that cold-climate animals tend to be large-bodied. Larger animals have less radiating surface area relative to their body mass, so they lose less heat and require less food, pound for pound. Marmots also have a cold-climate shape: round with minimal extremities — short legs, rounded ears, flattened muzzles. They're heavily furred for insulation. They spend all summer stuffing themselves with grass and wildflowers, storing those calories in a thick insulating layer of fat.

But a marmot's ace adaptation is its capacity to hibernate, a triumph of evolutionary engineering, like something out of science fiction. Marmots pass the winter in a state of suspended animation. Hibernating animals cannot be roused. Heart and respiration rate slow tremendously. Body temperature falls to within a few degrees of ambient temperature, just above freezing, to conserve energy. If you were trying to invent a strategy to get an animal through a long, impossibly harsh winter, it would be hard to conceive of a better device.

It's not easy to sleep a winter away. Marmots don't eat or drink for seven months, from late September to early May. Try it sometime. Human beings can't survive more than a few weeks without eating and no more than a few days without drinking. For one thing, we're poisoned by products of our own metabolism as fat and protein are broken down to provide for the body's immediate needs. It isn't entirely clear how marmots and other hibernators manage it.

The missing marmots of Marion Creek are a wonderful example of something else. They represent all the animals — caribou, pikas and, not least, grizzly bears — that could be here on this mountain, maybe should be here, but are not. The habitat could evidently support these animals. They prosper in similar habitats elsewhere. So why are they absent?

I've been working my way through David Quammen's excellent 1996 book, *The Song of the Dodo*. His topic: island biogeography in all its many-splendoured facets. Biogeography is the study of the facts and patterns of species distribution, the science concerned with where the animals are, where the plants are — and where they are not — and why, or why not. When the same sort of attention is focused on islands, it becomes island biogeography.

Quammen's book is a little over seven hundred pages long, stuffed with ideas and examples, but the essential principle for the nonce is this: An island generally contains less than its share of species diversity, relative to a nearby continent. An island harbours fewer species compared to an equivalent area of mainland habitat.

Certain kinds and groups of species are especially poorly represented on islands. Habitat specialists don't do well. Mammals, especially large-bodied, carnivorous mammals, are poorly represented on islands. This is probably related to another principle: The smaller the population of an animal, the greater the risk that something, even a run of bad luck, will push it into oblivion. Large-bodied mammals need a lot of space. Their populations tend to be small and widely scattered.

Smaller islands have fewer species. More remote islands have fewer species. In theory, over time, islands reach a biodiversity equilibrium, achieving a balance between extinction and the arrival of new species through immigration or evolution. But if an island is too remote for immigration, too small to provide the diversity of habitat and physical isolation required for evolution, then species are lost but not replaced. Biodiversity declines steadily, irreversibly. The term is "ecosystem decay."

The point to all this is that the word "island" doesn't just apply to a body of dry land surrounded by water. A mountaintop is the ecological equivalent of an island. An island in the sky. An island of tundra surrounded by a sea of forest.

About fourteen thousand years ago, not long in geological terms, much of North America was covered in glacial ice. In a band along the edge of that continental ice sheet, and extending much farther south along the cordillera, were large areas of tundra. When the glaciers retreated, so did the tundra, leaving only remnant patches on mountaintops — now isolated by more temperate ecosystems, much as rising post-glacial seawater isolated the islands of the Malay Archipelago from mainland Asia, or Tasmania from Australia, or Siberia from Alaska.

The alpine areas of Vancouver Island are doubly isolated. They are islands on an island, perhaps too small to support stable populations of large-bodied mammals over the long term and too remote for immigration. It's not surprising that the fauna here is impoverished. Nothing is more characteristic of insular ecosystems than the high risk of extinction. As Quammen says: "Islands are where species go to die."

Ah, well. Onward and upward.

Suddenly the summit seems attainable. To this point I've simply been moving on general principle toward higher ground. But now I can actually see a specific route ahead: so, so and so. The summit — to speak of it as a peak is stretching things a trifle — lies at the junction of four or five separate ridges. I'm sure a true mountaineer would just march straight to the top, *direttissima*. But the ultimate section looks a little steep for me. I decide to traverse one of the snowdrifts and come at the summit ridge, as I'm pleased to call it, from the north.

The snow traverse itself turns out to be somewhat tense. I have no ice-axe. Any attempt at self-arrest in the event of a fall is going to be iffy. The drift lies on a fair slope and drops away to, well, I know not what. I have some experience of just how quickly one can pick up speed after a fall on ice. So I proceed slowly, with great caution, kicking in steps. No problem. From there I face a steep climb over a talus slope of broken stone and a thickly vegetated summer meadow, a short scramble onto the north ridge and a stroll to the top.

The word "breathtaking" is vastly overused. In the literature of the tourism and hospitality industries, it's breathtaking this, breathtaking that and breathtaking the other thing. But from time to time, one does come upon a sudden view with the power to provoke an involuntary gasp: Ah! Wow!

It is indeed a grandstand view from the edge of Clayoquot Sound. The Marion Creek valley is a deep ditch below my feet, the slope more precipitous on this side of the crest. A flock of seagulls drifts by far below, white specks against green forest. The rugged pinnacles of the Mackenzie Range rise up on the far side of the valley. Between those peaks, I can catch glimpses of the Pacific Ocean, clear to the horizon in places, shining in the sun, otherwise obscured by fog or a low-lying cloud. To the north, the mountainous interior of Clayoquot Sound. To the east and south, the more gentle country of inland Vancouver Island. Farther east, in the hazy distance, the great snow-clad peaks of the Coast Ranges on mainland British Columbia.

You need a mountaintop perspective to fully appreciate the grandeur of a mountain landscape. The true ruggedness of alpine country cannot be properly gauged from below. The highest peaks are diminished by some trick of perspective or concealed entirely by lesser hills. From my high perch, I have a fresh view on a familiar world.

I wrote earlier that the influence of the Pacific Ocean is pervasive in Clayoquot Sound, impossible to escape, even up here. Alpine climate and conditions reflect the ocean. But it hadn't occurred to me that the mountain landscape might reflect the ocean in appearance also: dark curling waves of granite, a storm-tossed sea frozen in stone.

The view before me is a chaos of cirques, steep-walled valleys, knife-edged ridges, rocky spires — all excavated by ice. Once upon a time, before the ice ages, these mountains would have been substantially higher and more massive, but less rugged.

Interestingly, the highest peaks of the Mackenzie Range, opposite, are jagged and delicate as broken glass — not glaciated. At the height of the last ice age, those peaks would have been nunataks, islands of rock projecting above the ice field. Summits below about 1,400 metres are well eroded by ice. The Vancouver Island ice cap must have covered them all, rising almost to where I stand. I can see it in my mind's eye, a great swelling field of white falling away in a gentle slope toward the western ocean — clear in patches and glinting in the sun, otherwise shrouded in a layer of low cloud or fog, then as now. I could ski home, a long downhill run, except that home would be under the ice.

It's getting late. The afternoon is far advanced and I have a long walk back to the truck. But it's hard to go, hard to leave all this behind, hard to leave the realm of gods and return to mortal life. I've read, somewhere, that most mountaineering deaths occur during the trip down. Climbers linger too long on the summit. Then, on the descent, they run out of light or concentration or strength, and come to grief.

So, all this time while I've been blissing out on the mountaintop view, some part of my mind has been fretting about the trip down.

Worried about slipping somewhere and taking a tumble. Worried about losing the route and finding myself stranded at the brink of a precipice in the twilight. Worried about encountering a bear in the dark as I cross the clearcut. Worried about stepping into a hole and breaking my leg. I'm acutely mindful that if anything should go wrong, it will be a couple of days before anyone starts seriously looking for me. By then, of course, it will be far too late to do anything more than recover the body — if they can find it.

It's an oddly melancholy reaction to a joyful moment, but oftentimes these journeys into wilderness are an odd mixture of wonderment, euphoria and fear. It has to be so. One of the essential qualities of wild country, untamed country, is that it can kill you. This isn't from any sort of malicious intent, I don't believe that — though I do sometimes wonder, when I hear Windigo roaring through the forest — but simply from some ultimate indifference to our safety and well-being, campsites and paths that turn up in the nick of time notwithstanding. In wilderness we must take responsibility for ourselves. Or maybe: in wilderness, we're allowed to take responsibility for ourselves. A rare privilege in these times.

The trip down the mountain is slow, but relatively uneventful. I take a minor tumble on the heather. I stray once or twice from the route going down through the forest, little errors easily amended. Even so, the sun has set beyond the mountains by the time I reach the upper edge of the clearcut. Inside the forest, darkness is collecting, pooling rapidly in the hollows, obscuring the footing. It makes me glad that I still had some daylight left when I came down through the cliffs. Crossing the clearcut uses up the very last of the light: darkness has fallen by the time I reach the road. The clearcut also drains the last of my energy. My legs wobble and cramp as I head down the road toward the truck. That's okay, I have the sweet satisfaction of achievement to sustain me.

As I walk, I feel the darkness coming alive around me, in imagination and in fact. The night shift is waking up. My senses, too, are waking up. As the light fades, I hear and scent the world more clearly. I feel the crunch and roll of gravel beneath my boots. I'm more aware of the balance and posture of my own body, tired as it is.

As always with the coming of night, I feel a slow rising excitement that is not stricty fear, but the awareness of possibilities, a wild, dangerous excitement that is, at root, the contrast between vivid pleasure and high anxiety — the old atavistic terror of fang and claw waiting in the darkness. I am the beast and I have entered the game.

In the darkness I sense a sweet, close connection with this community of place and living things, an intimacy more intense than ever I'm allowed in the broad light of day, as if I have at last been admitted to the secret life of the place. Welcome. Beware. I feel a real pang when I finally arrive back at the truck, throw my pack into the cab, climb in and, all unthinking, turn the key in the ignition. Force of habit. The engine roars to life and the headlamps beat back the darkness, throwing up a wall between me and the night. Something precious flees before the brightness and the connection is broken, leaving me on the outside — again.

> *Farewell you northern hills, you mountains all goodbye*
> *Moorlands and stony ridges, crags and peaks goodbye*
> *Glyder Fach farewell, Cul Beag, Scafell, cloud bearing*
>   *Suilven*
> *Sun-warmed rock and the cold of Bleaklow's frozen sea*
> *The snow and the wind and the rain of hills and*
>   *mountains*
> *Days in the sun and the tempered wind and the air*
>   *like wine*
> *And you drink and you drink till you're drunk on the joy*
>   *of living*
>
> — Ewan MacColl, *Joy of Living*

September

HENIQOJÅSÅMIL

(Dog Salmon Moon)

Good weather lingers for an extra month and a half along the west coast of the island, in compensation perhaps for spring's reluctance or the foggy days of August, now past. While other places move toward brisk nights, frosty mornings and bright autumn leaves, these windward shores dream on through the extended summer of September and early October, oblivious to the passage of time. Alders are shedding their foliage, true enough, but that hardly counts. The leaves, still green, simply dry and drop, almost apologetically, as if not wanting to make a fuss: a faint, slithering rustle, like paper upon paper. There is no brilliant change in colour, no gold against blue skies. But the clock of heaven cannot be denied. Days grow shorter; dusk surprises us in early evening; the equinox is at hand. The hours of darkness multiply, hinting at our future. But the sky is blue, the sun still has some warmth in it and the ocean is summer calm.

In the old times, early fall would have seen the west coast people, at least those with household rights to a salmon stream, settled in their autumn villages ready to harvest the runs. Dog salmon, also called chum, were the single most important source of food for Native peoples. The fish arrived in early fall and hunters lay in wait with spears and traps; where salmon ran in great numbers, traps were more efficient. Traps were first set in tidal waters near the mouths of the rivers, and then later in the rivers *themselves. Clans with a productive stream might take thousands or even tens of thousands of fish in a few weeks. For the most part, men built and tended traps while women processed the catch.*

Nowadays, September marks a reluctant official end to summer. Kids go back to school after the Labour Day weekend and the number of visitors declines dramatically, though not so dramatically as it used to. Our secret is out, the travelling public have discovered the quiet days of early fall along the open coast. Nevertheless, Tofino is a different sort of town; gone is the August frenzy. Streets are quiet, parking is plentiful and lineups at our little supermarket have all but disappeared. There is a feeling, hard to shake, that the action has gone elsewhere. The rest of the world has moved on, leaving us to dream of endless summer.

Clayoquot River

WASTE NOT, WANT NAUGHT:
Coastal Temperate Rainforest

We are getting a head start on the September long weekend, the kayak and I. We have driven through the pre-dawn darkness, one gravel road after another, risking the shortcuts, taking curves too fast, all in hopes of catching sunrise at the narrows on Kennedy Lake. Now, as I loosen the tie-downs and prepare to unload my gear from the truck, the sun is just clearing the mountains. It's a beautiful blue-sky day, perfect paddling weather except for a faint high trace of cirrus cloud to the west, a suggestion of something brewing somewhere, still far off.

I'm going to pay my respects to the Clayoquot River valley. It may not have quite the same scale and grandeur as the bigger watersheds at the northwest end of Clayoquot Sound — the Sydney, the Megin, the Moyeha, the Bedwell — but among local people it has a reputation for extraordinary beauty. There is a certain Shangri-La quality to the place: a hidden wilderness beyond the mountains, lovely and untouched. This is the southernmost of Clayoquot Sound's large pristine watersheds. It may well be the southernmost pristine watershed — no roads, no logging, no development — of any significant size on the Pacific Coast of North America, everything farther south having been manhandled to a greater or lesser degree. That alone should make it worth a visit.

The valley takes its name from the Tła-o-qui-aht or Clayoquot First Nation, whose traditional territory it is. The Tła-o-qui-aht and their powerful chief Wickaninnish dominated this part of western Vancouver Island at the time of first contact with Europeans in the late 1700s, and the entire area came to be known as Clayoquot Sound, though other tribes held territory here as well.

I ease the loaded boat into shallow water, half lifting and half slid-ing, trying not to drag it across the sharp stones. Once properly afloat, I squeeze into the cockpit, fit the spray-skirt to the coaming and push off. A couple of good strokes and I'm away.

A wonderful day. The edge of the lake is rimmed with bright green reeds and shallows of clear water. Farther out, the deeper water is a vivid dark blue, already slightly ruffled by a breeze out of the west. Wavelets dance and sparkle in the morning sun.

The wider view is less attractive. The forests around the narrows have been heavily logged. Multiple clearcuts merge into one contin-uous ruin, up and down the steep slopes, along the ridgetops and out across the low country at the foot of the mountains. I'm not sorry to turn away and head north.

A familiar shape surfaces about fifty metres off my starboard bow, large liquid eyes watching me cautiously — a harbour seal that has found its way up the lower Kennedy River, chasing migrating salmon up into the lake. Kennedy Lake and the saltwater inlets of Clayoquot Sound are marvellously alike: the same steep rocky shorelines, the same little coves and beaches. I have to keep reminding myself that I'm paddling on fresh water. The likeness should come as no great surprise: it's a family resemblance. Kennedy Lake itself was once an inlet of the sea, excavated by glaciers flowing out of the mountains. But the slow rebound of the earth's crust after the immense weight of ice had gone gradually separated fresh water from tidewater. Now the only connection is via the lower Kennedy River. Like its sibling inlets, the lake is very deep. I read somewhere that the water at the bottom is still salty — a memento of earlier times.

I hug the lake's western shore, trying to get some shelter from the breeze, which is picking up more quickly than I'd hoped. When the steep slopes above the lake are warmed by the sun, the air next to them is also warmed. Warm air rises and cooler air floods in from the west to repeat the process: a chimney effect. Typically, early morning hours are relatively calm. There may even be a slight breeze blowing out of the high country, the process operating in reverse. By mid-morning the onshore wind will be starting to blow through the inlets and valleys. Generally the wind peaks in late afternoon and dies away through the evening. The process seems to be operating with a vengeance today. By the time I reach the head of the lake, whitecaps are blossoming all around and I'm looking forward to finding some shelter.

I've long since left the clearcuts behind. The scenery is a picture postcard of west coast wilderness: rugged mountains rising straight from the water, deep-cut valleys, steep slopes covered with dense forest. The narrow mouth of the valley is straight ahead and in the middle distance the highlands of the Clayoquot Plateau rise to bare rock and snow.

According to my map, the valley is quite broad except near its outlet, where a spur of rock looms from the east, creating the narrow gap through which the lower river emerges. There are two small lakes, Norgar and Clayoquot, the latter separated from Kennedy Lake by a short, steep stretch of river, the Lower Clayoquot.

I find the mouth of the lower river, which cuts a braided channel through the little estuary, and paddle cautiously over the bar into the current. Kayaks don't draw much water. You could float this boat on a heavy dew, as the expression goes, and I easily make way against the moderate current. The river is low and slow, as expected. It's a good sign. Plan A, kayaking up the lower river to Clayoquot Lake, still looks feasible.

I could hike from the estuary to the lake: Plan B. There is a path, but it would be a long tramp with all the gear. The route, mostly through scrub brush, doesn't seem very attractive. And I'd have to leave the kayak behind, stashed somewhere in the bush. Not my favoured option. I keep paddling.

For the first little stretch, all goes well. The river runs through a series of deep pools, quite lovely. But then I turn a corner and come face to face with disillusionment. Ahead, the river rises in a stepwise series of pools connected by abrupt, shallow riffles — clearly not navigable in any ordinary sense of the word. Reluctantly I abandon my hopes of paddling to the lake. But I remain unenthusiastic about backpacking. I extract myself from the cockpit and continue up the riverbed on foot, hauling the kayak behind me on a rope. Plan C, I suppose.

The sound of my precious boat grinding over the rocks makes me wince. The stones are padded and lubricated by a thick growth of algae, but there are still plenty of sharp exposed edges to gouge the fibreglass. That same algae makes for very slippery footing. Every so often my feet squirt out from under me and I go down cursing. Fortunately, the wetsuit affords some protection, I'm wearing my helmet and the water is relatively warm. So it's kind of fun. On this beautiful sunny day, the expedition has become a Huckleberry Finn adventure.

The bed of the river climbs more and more steeply until I'm lifting the kayak bodily over a mass of boulders tumbled from the mountain spur that looms above. I hardly care to think what this rock garden must look like under full winter flood conditions, when the river becomes an enormous rain gutter carrying Clayoquot Lake's overflow down toward the sea. On a lovely day like this, it's hard to believe the Jekyll and Hyde nature of these valleys. But I notice that even now there are little puffy clouds gathering all around the mountaintops: the air is rising up those steep slopes, cooling, condensing whatever moisture is present. The Clayoquot Valley, like every other valley on this coast, is a rain machine.

I have some interesting facts on this. A community-based research organization, the Clayoquot Biosphere Project, maintained a climate station in the valley for several years during the 1990s. Their equipment recorded an average annual precipitation of just under five and a half metres. The maximum monthly precipitation was 176 centimetres, and the maximum daily downpour was just over 30 centimetres. Not surprisingly, November and December were the wettest months, with October and January not far behind. The summer months were comparatively dry, but major rainstorms could occur at any time.

The most extreme flooding occurred in late winter, when rain fell on melting snowpack. On one memorable occasion, observers were forced to come and go via a makeshift bridge from higher ground. Their station, raised on pilings a good four or five metres above normal lake levels, barely escaped the cresting waters. Almost anywhere else, such flooding would be a catastrophe; here, it is routine, a normal occurrence. The combination of high precipitation, heavy runoff, steep terrain and unstable glacial deposits creates tremendous potential for runaway erosion, landslides and mass wasting. But the covering of forest, established bit by bit over millennia, maintains a surprising stability.

I heave the boat over the last big rocks and into the pool of deep water beyond. This is the lower end of Clayoquot Lake, really a short section of river impounded behind these boulders. I clamber thankfully into the cockpit and paddle to the middle of the pool where I can rest for a moment.

It is an extraordinarily beautiful spot. Sunlight slanting through the deep green water catches every detail of the round cobbles on the bottom. I notice a trout hanging motionless in the shadow of a half-buried snag. On either side of the pool, green forest gathers near, overhanging, illuminated around the edges by sunlight.

I paddle a few more strokes, amply rewarded by a lovely view onto the main body of the lake, framed by tall trees. A little farther and I'm into the open with a much wider view of the valley, completely forested and ringed with mountains.

The warmth of the day has provoked a swarm of little flies, which hover in milling clouds over the surface of the lake. Whenever one drifts too close to the surface, I see a quick splash, a hole opens in the water — and the fly is gone. A moment later I see the business re-enacted closer at hand. A school of little fish, none larger than my index finger, swim under the kayak. I notice dark vertical parr marks along their flanks; perhaps these are juvenile sockeye salmon doing their nursery year in fresh water. Even as I watch, a fly drifts too close. Quick as a wink, a little fish darts to the surface and snaps it out of the air.

The slopes along the east side of the lake are mostly exposed bedrock, supporting an open forest of shore pine, a hot, dry environment in the summer. On the west side, slopes are cooler and wetter, supporting a heavy growth of spruce, hemlock and western redcedar. About halfway along its length, the lake narrows abruptly. I land on the west side for a better look at the forest. I haul the kayak from the water and peel off my wetsuit; shirt and shorts seem more appropriate for hiking in this heat. Before I've finished changing, I'm thinking that mosquito repellent might be appropriate as well. Thoroughly bug-proofed, I set off to explore.

Like all ecosystems, temperate coastal rainforest is shaped by its physical environment. Ample rainfall and year-round mild temperatures favour a lush growth of vegetation, dominated by western hemlock, western redcedar, shore pine, yellow cedar, Sitka spruce and amabilis fir. This variety of species is a characteristic of old-growth forest. A wide variety of age classes is also typical — newly sprouted seedlings grow alongside very large, very old trees.

Wildfire is practically unknown here; there is too much moisture in the ground and in the vegetation. This favours the growth of large, long-lived trees like western redcedar and yellow cedar. The absence of fire also favours the accumulation of dead wood in the form of standing dead trees, called snags, and fallen logs in various stages of decay. This tremendous accumulation of biomass in living trees and decaying wood is another characteristic of temperate rainforests. In fact, massiveness is one thing that distinguishes temperate from tropical rainforests. A young Sitka spruce/western hemlock forest, for example, has about three times as much biomass as an equivalent area of tropical rainforest. Other forms are even more massive: the coastal redwood groves of California may outweigh tropical rainforest as much as ten to one.

When immigrant settlers began to arrive on the west coast of Vancouver Island in the late 1800s, they must have taken that massiveness as proof positive of the soil's fertility. I can imagine them thinking any ground that can grow such enormous trees will surely produce radishes the size of turnips and tomatoes like pumpkins. With backbreaking labour, they cleared land for gardens and farms, only to discover that the soil is surprisingly poor: mostly sand, clay and gravel, very acidic and lacking in soluble nutrients, which are leeched away by the heavy rainfall.

So what accounts for the luxuriant growth of this forest, which appears to be living on nothing more than air and water? In a couple of words: efficiency and thriftiness. When a big tree dies, much of its organic material is recovered. The continued lives of the forest, all future generations, depend on this recycling. That vast accumulation of dead wood is not a waste. It is, in fact, an essential store of nutrients, a savings account of insoluble organic material in the long process of being recycled. And how do living trees access this savings account? Well, they get by — and get high — with a little help from their friends.

A whole subcommunity of plants and animals — fungi, banana slugs, bacteria and so on — is devoted to the business of decay. They process the dead wood, little by little, turning it into products that living plants can use. Each downed log is home to billions of living organisms busily digesting its wood, and each supports a rich growth of vegetation living on the products of decay. Nearly every tree I see is rooted in the remains of a rotting log. A tree probably cannot attain full growth in this ecosystem without the benefit of one of these nurse logs. All through the forest, I encounter living palisades: rows of mature trees, their nurse logs long gone but still evidenced by straight-line cavities through the living root masses.

In return, living trees help to provide the conditions required by other members of the community. In the heat of summer, trees provide shelter from the sun and raise humidity by slowing the wind and collecting additional moisture from passing fog. In winter, they moderate temperatures by cutting the wind and slowing the radiation of heat from the ground. They also intercept much of the snow before it can reach the ground. And in the end, of course, each tree will die and lie down to nourish future generations of banana slugs and seedlings.

An ecosystem is more than a collection of individuals living in the same place. Each individual organism in this forest is supporting and being supported by other members of the community. The whole is greater than a simple sum of the parts. Isolated from one another, many of these plants and animals would die. Together they flourish.

It's worth observing that some shared benefit passes beyond the limits of this community. These trees are part of the planet's living machinery. They scrub carbon dioxide from the air and release oxygen. They moderate the climate upon which our human lives depend. And what have we done for them in return? How do we keep our part of the bargain?

The afternoon is well advanced now. Days are short in September, even if the weather is still whispering *summer*. Reluctantly, I call a halt to my walkabout. Time to get the tent up and see about some supper.

A couple of hours later, chores done, I take a mug of tea down to the lake, where I can sit and watch the evening light retreat up the eastern slopes. Clayoquot Lake is quiet and peaceful. A female merganser and her brood swim by, leaving barely a ripple.

The broad sweep of vegetation on the other side of the valley is a further lesson in temperate rainforest variety, on a larger scale. Old-growth rainforest is not a monotonous carpet of trees but a complex patchwork of habitat types, each patch home to its own mixture of species growing in response to some subtle variation in local conditions. Most slopes are dominated by the yellow-green foliage of western redcedar. But there are also daubs of darker green, perhaps amabilis fir or western white pine. Elsewhere, outcrops of south-facing rock support an open growth of shore pine.

In the absence of catastrophic wildfire, the scale of natural disturbance is relatively modest. Here and there, where windthrow or — more rarely — landslide has opened up a space in the canopy, I can see small, bright green patches of exposed shrubbery: salal, salmonberry and alder. Even the fiercest storms do surprisingly little damage to intact forest. In most instances of blowdown, just one or two trees are lost.

When an old tree goes down in a storm, it creates an opening in the canopy, restarting the successional clock. A sudden wealth of sunlight flooding the forest floor triggers a rapid growth of red alder — fast-growing, sun-loving, nitrogen-fixing — and other shrubbery. This rapid-response vegetation stabilizes and enriches the soil. It provides shelter from the wind and sun. It moderates temperatures. And it raises the humidity for shade-tolerant evergreen seedlings: hemlock, spruce and cedar.

Young alders are intolerant of shade and cannot grow beneath their elders. And mature alders are relatively short-lived. Soon they perish and release the well-grown young evergreens into full sunlight. The young evergreens grow rapidly, competing for space and resources. The weaker trees die. The strongest or luckiest individuals mature. In this way the forest is constantly turning over, tiny patch by tiny patch. Even within a particular habitat type we see a mosaic of differently aged patches, from early regeneration to ancient old growth. So the forest is both remarkably stable and constantly changing — albeit at a rather more leisurely pace than we're used to.

Someone sitting on this very spot, say, five hundred years ago, would have enjoyed virtually the same view. In fact, many of the trees

I'm looking at would already have been present as seedlings or young saplings. This summer evening in the Clayoquot River valley, this moment — the merganser swimming, the calm waters, the sunshine on the trees — could be from any one of the last seven or eight thousand years here. Think about it. Across all that huge span of time, almost since the last ice age, this community has sustained itself, day by day, in a more or less steady state. I suppose that's why these forests have such an immense air of timelessness about them. We can hardly help but be conscious of ancient and ongoing processes. And perhaps timelessness is another of the essential qualities of wilderness. There is an immutability to wilderness communities, a basic stability. Individuals may change, even neighbourhoods, but the general shape of the community endures, or changes only very slowly.

Human beings are so accustomed to the short term in our own affairs that we accept corruption and decline as the norm. Buildings fall apart, machines break down, nations and cultures go to pieces in the blink of an eye. All good things must end; nothing good can last.

But in places like the Clayoquot River valley we find biological communities that seem capable of functioning almost indefinitely: recycling nutrients, repairing themselves, making their own climates. This is not some far-fetched abstraction. This forest stands before me now, a tangible success. A tree beside my tent looks to be the better part of a thousand years old. And there is a space among the roots to memorialize the vanished presence of a large nurse tree, itself perhaps a thousand years alive and five hundred years dead when this living tree was a seedling. That's history you can reach out and touch — a visible and ongoing memory of the last twenty-five hundred years at this particular spot on planet Earth. This community has endured because it functions extraordinarily well. It is, by trial and error, perfectly adapted. Its longevity and continued survival are the proof of that.

I lie awake watching dawn come to the valley. I've been fretting a little about the weather. Someday soon, the first storm of winter is going to roll in. And I do not fancy trying to get down the Lower Clayoquot River after a rain. But the weather seems to be holding, thank goodness.

And there's a bonus: the mosquitoes are taking a break, at least temporarily. Either that or I now have insect repellent coursing through my veins. Cool air flows down through the watershed, whispering in the forest, bringing the fresh smell of high country. The sun clears the ridge across the lake, gradually illuminating the trees above me.

I discover an entirely new perspective, lying on my back, looking up into the canopy. The ground is up. The sky is down. I have become a fly on the ceiling of the cosmos. Trees seem to be hanging from the ground, dangling into the sky, straining toward the void. If it weren't for roots, they'd drop off and streak away like rockets. I roll over and *clunk*, it's all back the way it was. I remember hearing a painter talk about the phenomenon of defamiliarization, a familiar sight rendered strange and new. A crucial step in artistic perception.

It helps — if you're trying to reduce this riot of greenery to some sort of order — to think of forest as being structured like a layer cake. First, there's the mineral soil. It isn't easy to see the dirt beneath a temperate rainforest. The ground, for the most part, is covered with a layer of duff, decaying organic material, dead vegetation, bits of wood and so on, all stitched together with roots. But here beside my tent is a downed hemlock. The mass of roots has torn away from the ground and I can see the naked earth. Pretty feeble stuff, bare rock and gravel, nothing that we might call topsoil.

That mineral soil and the layer of duff are covered, sometimes almost concealed, by a thick growth of ground-hugging plants: mosses, ferns and succulents. Flowering plants on the forest floor are typically equipped with large green leaves and whitish flowers to take full advantage of whatever dim light reaches down through the canopy. Growing up through the ground-hugging plants are woody shrubs: salmonberries, thimbleberries, salal, huckleberries, blueberries. Towering above the shrubs are the stems and foliage of the huge trees that dominate this community. And finally, way up in the canopy, a whole subcommunity of plants and animals that spend their lives far above the ground.

Slowly, slowly, the edge of morning comes creeping through the canopy, down the great trunks until, at last, warm sunshine reaches the forest floor and my tent. The bright light catching the trees at such a low angle exaggerates the differences between species. The Sitka spruce in front of my tent is a straight, massive column supporting the canopy. The surface of the trunk is dark and textured with thick scales, rough as lava rock. Bright green moss, contrasting sharply with the bark, covers the trunk, except where chips of bark have come off to create a fresh surface.

If the spruce has a stony, mountainous appearance, the large cedar next to it looks more like living flesh. I wouldn't be surprised to see the whole trunk sway and flex like the limb of some great beast, muscle and blood. The bark is pale grey with vertical folds, like thick skin, pachyderm, or a covering of cloth. It supports less moss than the spruce.

Hemlocks are a bit nondescript by comparison. Being relatively short-lived, they're considerably less massive. The bark is lightly scored with vertical grooves. But they glory in their foliage, clouds of bright green needles, almost incandescent in the early morning sun.

Relieved of my weather anxieties, I am content to lie here for a moment, just watching. The biblical expression "a fullness of time" comes to mind; in a place like this you begin to have some inkling of what those words might mean. I wonder what it would be like to rest quietly here, in this one spot in the forest — on all the Earth — through a full late-summer day, allowing time to arrive like a river, bearing events to me. Slowly, slowly, the sun would rise high into the sky, move across the zenith and subside in the west. Shadows would arc across the ground, pool and fill the clearing to overflowing.

That would be the first day. But what about the next? And the next after that, week after week, fair weather and foul, year in and year out. A thousand years, say, rooted to this one spot. The life of a tree. After such a span of time, you'd know a thing or two about the forest.

Well, why not give it a try? Just for one day. I had planned to return home this afternoon, but I could put that off. I'll just lie here, watching, taking it all in. I could. But I'm a hasty creature, twitchy and restless. And my body is saying: *Not today, Jack, time to get up.* Besides, I want to go and see what's new on my lake.

The air over the water is alive with flying insects, all backlit by morning sunshine. The entire space is charged with vivid life. I can see more of the sky out here and there's not a cloud anywhere. I toy again with the idea of staying for another night, but decide not to press my luck. I'll go today.

I eat breakfast, pack and depart in leisurely fashion, returning the way I came over the rock garden, feeling relieved and grateful for such an easy passage. The water is even lower than it was the day before. Going with the current, the kayak doesn't need to be hauled. I can trail behind, nudging the boat every now and then to herd it in the right direction. Piece of cake.

Halfway down the lower river, I stop for a bite and a swim in one of the deeper pools. It's a hot day if you're wearing a wetsuit. Afterward, standing on the gravel watching the river, I notice, out of the corner of my eye, a flicker of movement upstream along the bank. Furry brown movement. Turns out to be a mink making its way down the river toward me. Completely absorbed in its foraging, the little animal, bright-eyed and bushy-tailed as the expression goes, runs between me and the kayak, an arm's length away, without so much as a sideways glance. On it goes, downriver, with that odd, looping, bounding gait of all the weasel clan — galoop, galoop, galoop — out of sight around the bend. Somehow I feel validated by the encounter, by the creature's confidence in me. Though I'd rather it was a little less trusting of strangers.

When I finally reach the mouth of the river, toward the end of the afternoon, the wind is still blowing hard. I take shelter on a little beach and wait patiently. Eventually the wind eases off. As soon as the whitecaps have subsided to a reasonable level, I slide the boat back into the water and head south.

Little by little, the breeze dies completely. The lake becomes perfectly calm, the surface of the water like a mirror. A squadron of loons, presumably two parents and a pair of their full-grown youngsters, rides at anchor, calling to one another.

There is no time to lose if I hope to make the narrows before the light goes. I paddle steadily — stroke, stroke, stroke — the kilometres falling away. The bow of the kayak surges forward, rippling into the mirror. It is one of those magic evenings when I feel in perfect tune with the boat. The power of each stroke seems to go directly into the lake. My paddle blades quiver, almost humming as they pass through the water. I'm reminded of the sound made by a large bird flying in still air, raven or eagle — stroke, stroke, stroke.

Now the sun edges toward the horizon. The water ahead, molten, reflects the sky, bending into ripples of gold and pale blue as I pass. The last light of the day retreats steadily, inexorably, up the eastern slopes of the valley until the final gleam rests upon the ridge — where it highlights kilometre after ragged kilometre of clearcut.

A familiar enough sight on Vancouver Island, and truth to tell, you get used to it. Those bare hillsides soon become just part of the scenery. Eventually you hardly notice them. Until you're fresh from a couple of days in the Clayoquot River valley.

It's another case of defamiliarization: a familiar sight seen from a fresh perspective. Suddenly in my mind, *clunk*, an image of the Clayoquot River valley stripped bare: my sense of the timelessness inherent in that place is gone, an illusion.

Ancient forests may seem all but immortal. And it's true that they are profoundly stable in the face of ordinary pressures. But they are exceedingly vulnerable to unaccustomed threats. I'm reminded of the many species of giant tortoise that once inhabited islands all across the south Pacific. They were immensely long-lived animals, wonderfully adapted to their environments, able to survive the most stringent of ordinary circumstances: drought and long periods with little or nothing to eat. But none of that careful adaptation could deliver them from gangs of hungry sailors, armed with axes and untroubled by any sense of obligation toward the rest of creation. Against unaccustomed threat, the great tortoises were defenceless. And doomed.

These vanished forests, like those in the Clayoquot valley, must once have seemed gorgeous and immortal. But to the men who demolished them, the trees were just another commodity to be cut, extracted, carted away — thousands of years of growth and slow development gone in a few months. And for what, I wonder? Newspaper? Disposable diapers? Two-by-fours?

There were no lessons learned on these slopes; this is no case of past mistakes, now bitterly regretted. The cutting only stopped because it had to: public outcry forced a reluctant halt. I don't doubt that some in the forest industry still see the delay as temporary. They're waiting for the fuss to die down, waiting for public attention to wander elsewhere so they can go back and finish the job. Certainly the industry will never, ever volunteer that the time has come to stop cutting. When we are down to the last few hectares of ancient forest, and we're getting close, they will still be demanding their "fair share" of whatever is left. They will keep cutting to the last great tree. And they will take that one, too, if they're allowed to get away with it.

But what a shame to destroy the last fragments of something so lovely and so rare; *eehmiss*, as the Nuu-chah-nulth people put it: very precious.

Compared to tropical rainforest, temperate rainforests have always suffered from exceedingly limited distribution. Even before extensive logging began, they covered just two-tenths of one percent of the planet's land surface, mostly slender coastal corridors along windward shores in the temperate latitudes. Most have now been cut. British Columbia and Alaska boast the lion's share of surviving old growth, but even here, most of the original rainforest has been logged.

Statistics from a mapping project by the Sierra Club of Canada illustrate the trend: approximately 2.3 million hectares of Vancouver Island were originally covered by coastal temperate rainforest. By 1954, about 30 percent of the old growth had been cut; by 1990, 64 percent. Put another way: more old-growth temperate rainforest was logged in the 36 years between 1954 and 1990 than in all the time previous. At the south end of the island, the rate of depletion was even higher: by 1990 over 85 percent of the original forest had been logged.

And, no great surprise, much of whatever old growth remained in 1990 is now gone. Vancouver Island boasts a total of 91 large primary watersheds, five thousand hectares or more. In 1990, 9 of those watersheds were still in pristine condition. As of this writing, only six remain intact.

And once that old growth is gone, it's gone forever. Even if industry is sincere in its promise to cultivate another crop of trees — which I doubt: it's hard to imagine the multinationals sticking around for 150 years, all that capital tied up, tending the crop, waiting for little trees to grow — and even if that crop of trees prospers, it will never be allowed to develop into anything like a natural forest. Wouldn't make economic sense. The industry envisions a uniform stand of trees that can be harvested at 100 years of age, just as they finish their rapid juvenile growth, like broiler chickens or feedlot heifers. Such managed forests will resemble wild old-growth forest no more than a cornfield resembles wild prairie grassland.

These are emotional issues. The kind of destruction that I'm looking at here on the slopes above Kennedy Lake always leaves me feeling angry and upset. It's not simply that something I value, something precious to me personally, has been destroyed. I'm as much offended by the callousness and disrespect toward the rest of creation, the arrogance, so eloquently bespoken by these ruined slopes. I do not say such things lightly. I have friends and relatives who work in the industry. I have no wish to offend them or anybody else, that's not my purpose or my style. I speak with all due respect. But it needs to be said: What happened here was wrong. No reasonable person can look at these mountainsides and conclude otherwise. What happened here should make everybody angry, including those who work in the industry, especially if they hope for some sort of future. There is nothing to take pride in here; it was not our finest hour. And I wish I could say this kind of logging no longer happens, but it does. Over much of British Columbia, clearcuts are still the standard approach to extracting timber. Slopes like this are being deforested as I write. What happened here still happens, all the time.

The farther I paddle, the angrier I become, thinking about the corporations and bureaucracies responsible. Have they no decency?

Well, no.

When I was younger, I took a certain amount of comfort in thinking that the most beautiful places at least would be spared. No decent person would harm them; they were simply too special to destroy. That seems unbelievably naïve now, but it was not an unreasonable thought back then, when companies in the forest industry were still proprietorships, with owners who had a sense of personal honour and a status to maintain in their communities. Such owners were restrained from doing anything really outrageous because their own reputations were at stake, and because, in those more benign times, the bottom line was not the almighty justification it is now. They might even do something grand to enhance their stature. But proprietorships are a thing of the past; it's all corporations now. And corporations aren't much hampered by any concept of beauty or decency or reverence; they are primitive organisms driven by a simple-minded appetite for profit. That's what they're bred for. Beauty and decency never make it into the calculation, unless they threaten the bottom line.

But there were real men and women involved here. Individual people. What about them? How could they bring themselves to participate? How did they talk their own good sense of right and wrong to sleep?

Life is complicated. People get swept up in the tide of events, often against their will. Jobs are scarce and the bills never stop. The men and women who work for the forest industry are mostly good folk who resent being cast as villains. Many have their own deep affection for the countryside. I know all that; I really do. But there are the clearcuts, the landslides, the ruined streams. Each big stump on these steep slopes represents a personal decision, an acquiescence. The need to make a living justifies a great deal in our culture, too much, I sometimes think, but it doesn't justify everything, not yet. We behave as if the fact of employment excuses us from our own moral and ethical obligations. *Mr. Just-doin'-me-job* can't be blamed, he's only following orders. Good people do awful things for the corporation or the ministry; for the job. But, no, we are not excused.

Four of the six large, intact primary watersheds remaining on Vancouver Island are protected, at least for the time being. That's good. But somehow I'm not comforted. Does that mean the other two survivors must be sacrificed, joining the 96 that industry has already mucked up? And what about places like the Clayoquot Valley: large intact fragments of much larger "developed" watersheds? Must they go, too?

So little is left. Surely the remaining old growth could be spared; the pristine valleys at least. It would be nice to see the right thing done for a change, the moral and ethical thing. Let the remaining fragments be spared, for decency's sake.

Things do seem to be moving slowly in that direction. There has even been some discussion around the idea of extending basic rights, a sort of legal personhood, to natural entities. An audacious concept, but no more so than other forms of emancipation must have seemed in their day (*What? Give women the vote? Are you mad?*). The word "emancipation" translates as "taking from the hand," a liberation from ownership. Legal history chronicles a steady enlargement of this basic franchise, an extension of legal rights to one group after another; a general shift in status from property to personhood. Why not? If we can grant the privileges of a legal entity with rights in a court of law to a business corporation, then could we not do the same for a river, a lake, a mountain, a forest, an ocean?

To see ancient forests as merely a source of economic opportunity is to trivialize them, and our relationship with them, beyond measure. Not everything in life comes down to the bottom line. Forests are a source of wonderment, enjoyment, peace and quiet. They are a welcome relief from urban landscapes and pressures. The ancestors of the Nuu-chah-nulth people found spiritual meaning in forests; surely the same is true of thoughtful people in our own times.

When I travel in wild country, I often feel the presence of something beyond my senses. I cannot tell if this is my own personal approach to the sacred or simply a dim perception of those realities that lie at the very edge of limited human consciousness and intellect. In any case, this awareness of the super-natural, the mysterious, is, to me, one of the most powerful attractions of ancient forest — of wilderness in general — and one of the strongest arguments for preserving it. I cannot hear those whisperings in urban surroundings. I am too distracted by the noise and the glare.

There is a strength that flows from undisturbed countryside, something that resonates in us, some sort of rhythm, a sense of rightness, perhaps. Always on these little journeys I am conscious of coming into tune with the natural world, of reawakening something that brings me energy and joy, as if the tide had turned in my favour. We are sustained in a real way by pristine wilderness. These are the landscapes of hopes and dreams, of endless possibility. When we destroy them, we cut ourselves off from a source of nourishment and strength. We are diminished. Wretched country begets wretched people. Ancient forests are part of the essence of life on the west coast

of Vancouver Island. We are a rainforest people, native or natural-
ized, by birth or by choice. These forests help to shape our sense of a
shared identity; of our own place in the world. What will become of
us when they are gone? We will become a people of strip malls, I suppose, and big-box retail outlets, and multi-lane highways to nowhere
worth going.

By the time I fetch the narrows, night has fallen. I let the kayak drift
to a stop at last, a hundred metres short of my final destination. I
pause for a moment, just resting in the darkness. The sky and the lake
still glow with a faint luminous blue but the shoreline is a silhouette
in black. I smell wood smoke. It drifts across the lake from the
forestry campsite, along with the insistent thump and thud of
duelling sound systems. An engine revs fiercely, once, twice, three
times. There's a sudden sound of truck tires chewing on gravel.
Headlights brush the darkness, moving rapidly in the direction of
town, disappearing around the bend. The commotion fades. A bank
of dark clouds hangs in the west, a change in the weather right on
schedule. Even so, Jupiter shines like a diamond in the clear air
above. I take up my paddle reluctantly and head for shore, bracing for
re-entry.

> *Courage my friends; 'Tis never too late to make a better
> world.*
>
> — Tommy Douglas

October

ETSOSIMIL
(Rough Sea Moon)

O *Winter comes at last in October. The month begins* *with a lingering spell of good weather. But conditions change abruptly* *around the Thanksgiving weekend, as if someone, somewhere, had thrown* *a switch to send a great ponderous mechanism revolving into action. The* *first storm blows in and summer never quite recovers. Soon comes another* *disturbance and another as the storm track sags southward. Even between* *disturbances, the air is different. Gone are the skies of summer, cloudless* *from horizon to horizon, hazy with humidity. Now the mountains are* *sharp and clear in outline; patches of brilliant blue sky play hide-and-seek* *between towers of great billowing clouds.*

In the old times, that change of weather would have seen the west coast *people still hard at work along the rivers, harvesting salmon. Catching fish* *was the least part of the job. Each salmon had to be cleaned and filleted, the* *flesh dried and smoked, an enormous labour that went on throughout the* *fall. Fish were cut along the back and the flesh carved off the bones toward* *the belly. That left the head, tail and backbone in one piece and the flesh in* *another: one wide strip, the two fillets connected at the belly. Once dried* *and smoked, the flesh could be packed in bales for storage. Autumn was* *also the time to gather roots of wild clover, Pacific cinquefoil and bracken* *fern (now thought to be carcinogenic and not recommended as food).*

In modern times, October is a source of mixed feelings. Like some sweet dreamer waking abruptly to discover that he's overslept again, we come dazed and startled to the end of our extended summer. The calendar says we are a month into the fall, with winter just around the corner. Brutal. There is a natural nostalgia for sunny summer weather. Nobody looks forward to getting cold and wet. On the other hand, winter is traditionally a time for building up the fire and settling in, visiting, reading, making music; a not altogether unpleasant prospect.

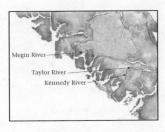

Megin River
Taylor River
Kennedy River

RING OF BRIGHT WATER:
Bears, Salmon and Wild Rivers

Hishuk Is Ts'awalk: Everything is One

Early one blue-sky summer morning on the farm, when the world was still young and fresh, my grandfather and I took leave of my grandmother in her kitchen. He ushered my small self outside into the sunshine. She watched us go, waved goodbye and closed the screen door. We two men walked together across the lawn to the bottom end of the yard where we could look down over the cutbank, onto the slow-rolling river that was the farm's southern boundary. After a moment, we turned and headed upstream along the bank, moving easily with the sun on our faces.

Half an hour later found us sitting quietly, half-hidden, in tall grasses still wet with dew. Mist rose from the river in the growing warmth of the day. Set back from the shore, an abandoned orchard lay tucked in against the dark forest. Fifty or sixty feet away, the object of our attention — a large adult black bear — foraged for green windfall apples that had accumulated beneath the gnarled trees.

That bear was aware of us, knew it was being watched, but continued to feed, snuffling in the grass. Long fur, coarse and midnight black, shone and quivered in the light. We talked in whispers, my grandfather and I.

Then the animal paused, looked up, raised its snout to quest uneasily back and forth, trying to recover the faint trace of some odour. Perhaps it had scented one of the dogs back at the house.

Suddenly it wheeled, plunged noisily off through the grass and vanished into the bush.

We never saw it again. Doubtless that bear is long gone to bones and green grass. The farm is gone. My grandfather and my grandmother, too; all taken by time. But the memory stands and is before me yet. It is part of the man I have become and still ripples all down the long years to some unknowable future effect.

Shelter Inlet. A fish the size of a porpoise torpedoes from the water and splashes back. *Krsplash.* Ripples spread just ahead of the kayak. It could only be an adult chinook salmon, very likely hatched from one of the rivers nearby. It will have spent the last four or five years along the coast of British Columbia and Alaska, hunting smaller fish and being hunted in turn by sea lions and killer whales. Now it has returned, fully mature, to spawn in its natal stream.

But the weather is not cooperating. The last couple of weeks, late September and early October, have given us nothing but blue sky and sunshine. The autumn rains have yet to arrive and there is a dearth of water. The big fish must bide its time, waiting.

Krsplash. And again. *Krsplash.* Silence. I wonder if that fish would be so eager knowing that this journey home will be the death of it.

I rest in calm water, planning my next move. The warm light of late afternoon colours the forest along the shore and the steeper slopes at the head of the inlet. From somewhere off to the right comes the sound of a waterfall. All quite lovely, but I'm tired. It's been a long day — seven hours of steady paddling from Tofino to the mouth of the Megin River — and it is not over yet. I must find a place to camp and darkness comes early at this time of year.

The Megin River valley is the largest undeveloped watershed on Vancouver Island. A relatively low-gradient stream, it is home to an exceptionally diverse biological community. Adrian Dorst — who gets around more than most — once told me that Megin Lake was as pretty a place as he'd seen in Clayoquot Sound. I've been wanting to see it for myself. I'm going to spend tomorrow on the river. I'll go as far upstream as I can; with a little luck I'll make it all the way to the lake.

The Megin empties into Shelter Inlet through a little bedrock gorge. I point my bow into the current and paddle hard, working my way slowly upriver. I enter, and will later leave the Megin, on a high tide, pure chance both times. Weeks later, Dan Lewis — Daniel-San, the kayak-master — tells me that the mouth of the Megin is a waterfall at low tide. Oh? I could just imagine paddling down the river, oblivious, with the kayak fully loaded. Somebody's looking after me.

There's very little water in the river. The kayak grounds on the gravel before I've gone a hundred metres. Perhaps that's just as well. Evening is coming on and I desperately need to find myself a campsite. I haul out and spend the next half hour scouting up and down the banks. Eventually I find a little spot back among the big trees. It's even darker under the canopy; I have trouble seeing well enough to set up the tent. But at least I'm out of the riverbed. Safer up here, I'm thinking. Not that it's very likely the stream would rise tonight, though such things do happen; I'm more concerned about the possibility of a midnight visitor. While unloading the kayak, I happen to glance upstream and can just discern a large black shadow on a gravel bar up there. A bear? I watch for a long time. There's no movement. Of course, he's probably thinking the same thing.

Perhaps it's also just as well that the salmon haven't entered the river. It might be a little nerve-wracking to camp in the midst of a teddy bears' picnic. And what would I do for drinking water with the Megin full of dead salmon?

I'm too tired to cook. It's dark and I'm still nervous about that shadow on the gravel bar. I make do with what's left of lunch and get to bed.

"Hishuk is ts'awalk" is a Nuu-chah-nulth expression that comes up time and again in conversations about the environment in Clayoquot Sound. It means "everything is one" and water is the prime example. The whole country is a spreading network of streams and rivers: a mosaic of watersheds, catchment basins, drainages. Waterways are the landscape's circulatory system and its basic unit of organization.

From sky to sea — mountain tarn, ephemeral trickle, stream, spreading river, estuary, inlet, open ocean — water is the tie that binds. From visible surface waters to the hidden gravels of the hyporheos and ultimately into the dark, cold stillness of the phreatos, Clayoquot Sound is one vast moving body of water.

The greatest terrestrial diversity and productivity are found along waterways. Streamside or "riparian" habitats are a magnet for complex plant and animal communities. Surface waters and their adjacent habitats are intimately linked, so much so that they are more properly thought of as a single ecological entity: the "hydroriparian ecosystem." In each watershed, networks of riparian habitats comprise a continuously connected environment that provides for movement and dispersal of animals and plants throughout the landscape, including passage over the ridgelines and into adjacent watersheds. *Hishuk is ts'awalk; Everything is one.*

In the morning I'm stiff from yesterday's long paddle, but looking forward to my walk. Remembering my experience in the Clayoquot Valley, I've decided to leave the kayak behind and proceed on foot up the river's empty bed. With the water so low it should be a simple matter. Fortunately I thought to bring my rubber sandals along. But first some breakfast. The sun is a long time getting clear of the trees, so I carry stove, food and paraphernalia out onto the gravel bar and set up my kitchen there.

The stew is coming nicely when I hear rock against rock and look up to see a small black bear coming upstream along the far bank. The river, what little there is, runs between us. He or she is still a good way off and may not even know I'm here. Presumably, like that big fish last night, this bear is just hanging around, putting in time until the run begins. I stand and shout. The thing is to avoid surprises at close quarters. To my consternation, the animal keeps coming. I move farther out onto the bar, very domineering and obvious. I pick up a couple of rocks and bang them together, hard, the loudest sound I can make, and shout again. Impossible that the creature has not heard me, but still it keeps coming, a steady pace, neither faster nor slower. I begin to wonder if, somehow, this bear is habituated to human beings. The Megin sees a fair amount of traffic in the summer and I suppose that not everyone is as careful with their garbage as might be. There isn't much more I can do except to watch it come.

In the end, the bear travels the length of the far shore, past me and out of sight upstream. Never even glances my way. Maybe it was simply doing me the honour of treating me like another bear. It could not afford to retreat from this soon-to-be rich feeding station. It must assert its right to be here; survival depends on that — to flee is to die — but it was also very circumspect, very focused on its own business, not antagonizing me at all.

Bears are solitary creatures for the most part, quick to flee even from their own kind, most comfortable alone. But in autumn they congregate along salmon rivers, drawn together by the windfall of food. For them this is a very, very tense social situation, the close quarters almost unbearable, so to speak. But crowding must be borne because they cannot afford to neglect a prime source of calories just before winter. So they follow an etiquette of sorts that allows them to share the food without antagonizing one another to the point where someone might get hurt.

I feel a bit silly making such a fuss. I know it was only prudent, by the book, and I would do it again. Still, I feel a bit like an offensive neighbour, loud and obnoxious, too protective of my own little bit of turf. And I was so absorbed in risk management that I forgot to enjoy the experience, a close encounter of the best kind, where nobody gets hurt.

After breakfast, everything carefully washed and hung high out of harm's way, I head up the river. Hard to think of a more pleasant way to spend a day, splashing across the shallow riffles, wading the clear pools. The water is not too cold and the autumn sun is bright and warm. Here and there I find myself pushed into the forest by deep water stretching bank to bank. There are patches of devil's club back there I'm not crazy about. Hard on the bare legs.

There is a family resemblance among rivers and a repetitiveness within rivers that is like rhythm in music. If rhythm in music is an echo of the human pulse, rhythm in a river echoes the pulse of water: the hydraulic jump, lift and drop. Bar-riffle-pool, it goes. Bar-riffle-pool. Curve left, scour pool. Curve right, scour pool. Bar-riffle. Bar-riffle. The repetitiveness is not unpleasant, with variations, and it has the charm of a pattern recognized. Every now and again comes a cadenza to arrest the stream's headlong flight. Water collects in deep still pools behind the natural weirs of rock or wood — usually great logs buried in the bed of the river — before plunging down to begin the next passage.

From the naturalist's point of view, each distinct set of conditions, each phrase and measure, creates a discrete pocket of habitat required by a specific community of living things. Or to put it the other way: each different set of physical conditions shapes a different sort of living community. Call it counterpoint or harmony. Embellishment.

The run of the river doesn't stop at bank or bed. Water continues to flow, albeit more slowly, through the interstices of gravel, soil and even porous rock. These "hyporheic" waters have their own biota:

minute creatures living their lives in this under-river. The hyporheos is a transition, an ecotone, between the surface stream and that vast, slow pool of groundwater proper, the "phreatos." There is a constant exchange of water, nutrients and biota between the stream, the hyporheos and the phreatos. The free-flowing surface water is just the most visible component of a much larger waterway, the tip of the iceberg. *Hishuk is ts'awalk; Everything is one.*

Interactions and interrelationships extend above ground as well. The forest is an important influence on the stream, fundamentally shaping its physical form. Streams in old-growth temperate rainforest are characterized by what hydrologists call "large woody debris": enormous logs, the remains of mature trees, which fall into the stream and become partly buried in sand and gravel, causing the stream to meander and cascade, creating pools, backwaters and a variety of habitats critical to fish and other aquatic creatures. The buried wood also slows and captures the gravel, rocks and sediment moving along the bed of the stream. In countryside that has been cleared of forest, streams lose their large woody debris and blow out, becoming crude ditches of gravel and rock, straight and sterile.

The forest also influences the stream in less obvious ways. Bankside vegetation moderates water temperature, providing shade in summer and shelter in winter. And the forest is an important source of nutrients for the stream: aquatic creatures depend on insects and bits of vegetation that fall into the water.

The forest also moderates the effect of torrential rains. The thatch of organic matter on the ground armours it against the downpour and old-growth forest soaks up the excess moisture, releasing it gradually. Streams like the Megin may rise and fall dramatically during a storm, but they rarely become catastrophic muddy torrents, which is astonishing given the amount of precipitation and considering that much of the countryside is buried in glacial till. All this is critically important for the salmon that spawn here. Silt settling into the gravel, the hyporheos, would slow the circulation of water and suffocate eggs.

Benefits also flow the other way. A group from the University of Victoria under Dr. Tom Reimchen (aka Stickleback Tom) has collected fascinating data to demonstrate the relationship. Reimchen became interested when he noticed that forests bordering salmon-spawning streams were littered with partly eaten fish carcasses, thousands of them. Subsequent investigation suggested that 80 percent of the salmon spawning in a small stream may end up on the forest floor, carried there by predators and scavengers, particularly bears.

It happens that salmon carcasses contain a significantly higher level of N^{15}, a heavy isotope of nitrogen, than you would normally expect to find in terrestrial biomass. By measuring the relative levels of N^{15} in forests surrounding salmon-spawning streams, Reimchen and his group demonstrated that a substantial portion of the nitrogen going into those forests is marine nitrogen, mostly from salmon carcasses. In portions of watersheds where salmon cannot spawn — above waterfalls, for example — N^{15} levels are dramatically lower, equivalent to norms elsewhere.

It has long been supposed that the annual windfall of salmon carcasses is an important source of nutrient for streams and rivers. The dead adult salmon enhance the fertility of their natal stream to the benefit of their own progeny. But now it appears that salmon are also fertilizing the surrounding forest. It is no coincidence that salmon streams like the Megin wander through forests of magnificent trees. The stream and its salmon built those trees. And when the trees have finished growing they will fall into the stream to provide habitat for eggs, fry, smolts and spawning adults, which in turn provide food for bears. *Hishuk is ts'awalk; Everything is one.*

I see no bears today. But there are tracks everywhere and a strong sense of presence. They are here, they just haven't shown themselves. *If you go out in the woods today, you're in for a big surprise ...*

I'm not going to make the lake. It's just too far and the afternoon is already wearing on. The deep pools and bushwhacking have slowed me down. It doesn't matter. I have what I came for: a sense of the river running. So I give up; a little too easily, perhaps. I'll keep Megin Lake as a treat for another day, an abstract icon of beauty, unattainable for now. I do an about-face and start downstream, back the way I came.

Here's the rub, then: as long as the weather is good, I'm not going to see any spawning salmon. Simple as that. The salmon need water in their rivers and there's not going to be any water until we get a storm. But once the weather turns, it will become impossible for me to visit wild rivers in remote corners of the sound. I'll have to settle for something less pristine, more accessible.

The sky is still overcast when I leave Tofino, but at least the pavement is dry. The expected change of seasons clicked in a week ago, mid-October, right on schedule. The most recent disturbance went through day before yesterday, not a major storm but enough to fill the rivers. I didn't bother going out. It would have been hard to see the fish with so much water coming down, not to mention danger-ous. But the salmon will have taken advantage of that surge to move upstream. Friends report good numbers of fish in both the Kennedy and Taylor rivers.

The drive up the Kennedy River is a little dreary with clouds hang-ing low in the valley, but at least it's not raining. The alders are almost bare of leaves now. These trees are growing on old clearcuts, second growth, an unhappy association, but they have an undeniable beauty, especially at this time of the year when their gleaming white trunks stand out so vividly against the darker forest beyond.

I'm planning to drive all the way over Sutton Pass into the Taylor River Valley. That watershed has also seen a good deal of disturbance, but access is easy and my chances of finding a few spawning salmon are fairly good. On the way home I can check out some places on the Kennedy River. It won't be quite the same as visiting the Megin, but I should see some fish.

Alongside the Taylor River, I turn off the pavement. Two or three kilometres on a gravel side-road brings me to a place where I can park beside the water. The day is brighter on this side of the pass, there's even a bit of wan sunshine coming through the clouds, lighting what's left of the autumn foliage, mostly bigleaf maples. Even before I leave the truck I'm seeing dark crimson-purple patches out in the river.

These are sockeye salmon (*Oncorynchus nerka*), masses of them hanging in the deeper channels. Spaced all across the shallow gravel bar are pairs of mating fish. In the ocean, sockeye are beautiful streamlined torpedoes, white below, silver sided, bluish-black above — some fishermen call them bluebacks. Here in the river, both males and females are a brilliant red, except for their pale green heads. Both sexes have the hooked snout from which the genus takes its name, though the trait is more highly developed in males. Males also develop exaggerated teeth, a prominent humped back and uptilted head. The lovers are both grotesquely beautiful and beautifully grotesque.

It's the female who excavates a nest, the "redd," in gravel. She manages this by lying on her side against the bottom and flexing her tail vigorously upward. The gravel is lifted out by suction. I'm surprised at the coarseness of the gravel where these fish have chosen to nest. Some of the individual stones are not much smaller than my fist. The force of the female's digging must be considerable and it must be hard on what's left of her body.

The male stands guard, which is to say he chases away other males. As the nest approaches completion, the female stops occasionally to test it with her ventral fins. The male moves alongside, touching her flanks gently with his snout. Sometimes he quivers rapidly next to her. Then he'll cross over and quiver on the other side. When the female is satisfied that the nest is deep enough, she will crouch in the cavity and, quivering, release her eggs. At the same time the male, also quivering, releases his milt, a cloud of sperm to cover the eggs. Occasionally a second male will dart in on the female's other side and release his milt. The eggs are slightly sticky when first laid, so they adhere to the bottom of the nest. Almost immediately the female moves upstream and begins digging gravel out of the bottom just ahead of the nest, to bury the fertilized eggs. This second excavation also becomes a nest and the process repeats. The female may dig and fill as many as seven nests, most commonly four or five.

And then she dies.

The male lasts perhaps a few days longer, fertilizing the eggs of other females. Then he dies too.

The rate of fertilization can be surprisingly high, approaching 100 percent. Fertilization must occur very quickly after the eggs are laid, generally within 30 seconds. After a single sperm enters the micropyle canal in the egg capsule, other sperm are prevented from entering. Fusion of the two haploid nuclei occurs within an hour, the first cell division within 12 to 18 hours. As soon as the eggs are shed, they begin to absorb water. They swell and their internal pressure rises. Within a few moments of water activation, the eggs become exquisitely sensitive to movement or shock. The slightest jarring can kill them.

These eggs will hatch in about 60 days. The tiny "alevins" will stay in the gravel until spring, living off their remaining yolk. The eggs and alevins require an adequate flow of water through the gravel. Siltation can cause high mortality. Oxygen consumption increases after hatching and alevins sometimes move through the gravel in search of better circulation. When the yolk is almost used up the young salmon, now known as "fry," emerge from the gravel and swim downstream.

From the sockeye point of view, this stretch of river is prime real estate. Young sockeye spend from one to three years — mostly just one year — living in freshwater lakes. The nursery for these fish is Sproat Lake, just a few hundred metres downstream.

Fry migration can be very complex. For instance, one researcher described sockeye fry emerging from Weaver Creek in the lower Fraser River valley on mainland British Columbia. That creek flows into Morris Slough, which empties into the Harrison River midway between Harrison Lake and the Fraser. Fry emerging from the gravel of the creek are carried downstream into the slough. Then they swim downstream about one kilometre to the Harrison River where — and this is the amazing bit — they reverse their orientation to current and migrate *upstream* about five kilometres into Harrison Lake. And they carry out the whole manoeuvre with such dispatch that one might think they'd all been handed a map and timetable before starting.

In their second spring, a year after emerging from the gravel, most of the fry from these eggs will transmogrify into silvery "smolts" with all the physiological equipment necessary to make the dramatic change from fresh to salt water. One dark night they'll be off, downstream through the lower Sproat River into the Somass River, down the Somass into Alberni Canal, out the canal, through Barkley Sound and into the open ocean for two to four years of deep-water wandering.

And then, sometime in the summer of their fourth year, when fish from this river are scattered all across the eastern half of the north Pacific, it will be as if the call had gone out. As one, they will turn and make a beeline for Barkley Sound, moving almost as fast as they are able. They will arrive there together and simultaneously, as if for a scheduled meeting. Travelling as a group, they'll ascend the Somass River, the Sproat River, Sproat Lake and ultimately the Taylor River, to "Begin the Beguine," again.

Opinions differ, but most experts believe that salmon have evolved from freshwater ancestors. Somehow the group developed the necessary physiological equipment and an inclination to spend some portion of their life cycle in salt water. It's a very effective strategy. The ocean offers a richer food supply than cold, freshwater rivers and lakes. But rivers may offer some sort of advantage for developing eggs and fry: freedom from predation perhaps. Or it may just be that salmon haven't quite been able to shake that last link with their past.

If the theory is correct, those species that spend the least time in fresh water, chum salmon (*Oncorhynchus keta*) and pink salmon (*Oncorhynchus gorbuscha*), have evolved farther away from the ancestral stock. Some pink and chum stocks can actually spawn in the intertidal. Species that spend more time in fresh water, coho (*Oncorhynchus kisutch*) and sockeye salmon (*Oncorhynchus nerka*), are less highly evolved. Chinook (*Oncorhynchus tshawytscha*) are in a class by themselves, midway between the other species. Some chinook stocks spend a year or two living in fresh water, while others take to the sea right after hatching.

What is it about the sight of salmon in the river that I find so deeply moving? Certainly few other experiences speak so clearly of, are so powerfully charged with, the flavour of autumn in wild country on Canada's Pacific coast. But I think the fish in the river also whisper something essential about the very nature of life. As writer Daniel Wood once observed, such encounters provide a window beyond the actual facts onto things fundamental. They allow us to glimpse, however briefly, something enduring and profound.

I sit for a long time, just watching.

Then I climb back into the truck and head for home. Once more in the Kennedy watershed, I pull onto another old logging road that delivers me to a bridge locally known as Jump-Off Bridge, spanning a deep pool of the river. I want to try something new and different — for me anyway. I pull on my wetsuit and carry my snorkelling gear down to the water's edge.

Oh my, but it's cold! I can't remember swimming in colder water. I cross the pool, nervously alert for any glimpse of bear, for animated shadow moving through the brush or across one of the gravel bars. Much as I like bears, I don't want their company at this moment.

There is an interesting line of speculation that says the evolutionary or adaptive reason for bright colours and grotesquely enlarged bodies in spawning male salmon is that it makes them more attractive targets for stream-side predators, which, thus distracted, allow the nondescript but egg-laden females to slip through relatively unscathed. And research suggests that, indeed, a significantly higher proportion of fish taken by bears are males.

Doug Palfrey — senior surfing dude and manager of the Tofino

Salmon Enhancement Society hatchery — talks of having a bear come for him once during a swim. It happened when he was surveying a stream for spawning salmon. The bear evidently took him for the great grand-daddy of all *Oncorhynchus tshawytscha*. Bears are notoriously short-sighted; Doug doesn't look the least bit like a chinook, especially a spawning chinook. Fortunately his dogs managed to drive it off.

I like to picture that bear, later, talking with his buddies. Miller time.

You should have seen the one that got away today. Thing must have been six feet long. A monster. Damn dogs anyway. I saw it first.

My face is so cold it *hurts*. But I'm about to see a sight that's worth a little pain. At the bottom of the pool where the water accelerates toward the downstream riffle, I run into a school of perhaps a hundred salmon. And I see them from a fantastic new perspective.

Usually we see fish from above. And we see them in just two dimensions, like an animated photograph. Things are moving around, but it all happens in one plane, like television. These fish are in three dimensions. They hang in the current like birds in a wind. And they're *watching* me. Their eyes are on me and I can see they're wondering what the hell I'm up to. As I move toward them, half out of control, like a rogue dirigible, they smoothly part ranks to let me through, evidently confident they could outswim this monster should the need arise. They have an aura of muscular compactness, power and strength. There are other fish in the water, cut-throat trout I guess, waiting for eggs.

Too soon, the current carries me past. Actually, I have no sense of being carried along. It seems rather as if the green cobbles on the bottom are sliding past me, moving upstream, accelerating, like a train moving out of the station. It's an odd sensation. Through the water I can hear the roar of turbulence downstream.

I repeat the swim two or three times, then head back to the truck, shaking with cold.

These are also sockeye, though they don't seem quite so far advanced toward spawning as the fish in the Taylor River. Not a very big school, either, bit of a disappointment. Once upon a time the

Kennedy River boasted one of the largest runs on the west coast: hundreds of thousands of fish. The river would have been a mass of red. I wish I'd seen it: it must have been one of the world's wonders. But during the 1960s there was a series of intense gill-net fisheries at the mouth of the river, and the Kennedy watershed was heavily logged. A man once told me of watching heavy equipment working *in* the river, of seeing pink salmon eggs in the gravel of logging roads. And that was that. The runs have never recovered.

Back on the highway, heading home, I pass a sign announcing THORNTON CREEK SALMON ENHANCEMENT SOCIETY. I don't give it much thought. But when I reach the Tofino–Ucluelet junction, I yield to impulse and turn left toward the village of Ucluelet.

The Thornton Creek hatchery is on the east side of Ucluelet harbour, opposite the town, just back from the mouth of the creek. It's a small cluster of buildings and rearing tanks, all surrounded by an electrified chainlink fence, not so much to protect the property from people as from bears. It's just a little hatchery, nothing like the scale of the big Canadian government hatcheries in Port Alberni and Vancouver; a local effort, mostly run by volunteers who have long since gone home for the day. But they've built a boardwalk along the lower section of the creek, to accommodate people like me who come down to gawk.

This lower section of the creek is tidal, high water just now, and full of huge fish. These are not sockeye. These are chum salmon, *big*, and chinook salmon, *bigger*. Neither has the brilliant red of the sockeye. Chum have a patchy calico pattern of purple and green, teeth very prominent in the males. Chinook are mostly green with a wash of red along the flanks.

I would guess that pretty much all of the fish in this pool are products of the hatchery, an artificial run entirely dependent on enhancement. Thornton Creek couldn't support much of a run on its own. It's just a little creek, not much volume, and only the first hundred metres are accessible to fish. An abrupt slot in the rock with water cascading down from above absolutely blocks any further passage upstream.

As an experience of salmon and moving waters, Thornton Creek seems about as different from the Megin as can possibly be. The whole establishment has a tidy, ship-shape, business-like feel to it, faintly industrial, reminiscent of agri-business facilities I've visited: greenhouses, chicken hatcheries, that sort of thing. At first glance, you wouldn't think it could pack much of an emotional charge.

So I'm altogether unprepared for what I find at the far end of the boardwalk. The deep pool at the base of the waterfall is full of big fish, ragged and dying, circling slowly, circling and circling, waiting. These very fish have crossed whole oceans, driven by their own restless urges, frantic to be here, not wanting to be late. Not wanting to miss out. Now they have made it. Now they are here. And I wonder: do they have some sort of hope or expectation in their dim, fishy minds? What are they waiting for down there? Revelation? Ecstasy? Or just Deliverance?

Here is the bitter end of things. All that far-flung wandering across the wide oceans under the stars, and it comes down to this dark little hole in the rock. They can't even see the sky. I am reminded of my grandfather in the bedlam of extended care, a terrible place made worse by the tacit understanding that — like this pool — it was the very last place: no getting out, no going on, no happily ever after. They're all just waiting for death to set them free.

It all seems so pathetic, so futile, so unjust.

This is no easy thing, coming to terms with the ugly fact of mortality — my own mortality, certainly, but especially the mortality of loved ones. Like most people, I shrink from the fact of death, don't want to deal with it. Death is inevitable, we tell each other casually, can't be helped. Death and taxes: isn't that the expression? Cold comfort. Death comes to all of us living. If we want to be alive — if we plan to fall in love with living things — then we'd best get used to the idea. Death is part of the deal, take it or leave it.

But that's not exactly right, either. It's not as if we've hammered out some sort of bargain with God; that death is the price we pay for being allowed to live, the quid pro quo, a tariff imposed. Not at all. Better to think of it as part of the process of living, of being alive.

Death is the natural consequence of life, not its bane. Death is not foreordained. Barring trauma or disease, we don't die all at once, like an engine running out of gas, like a light switched off. We die little by little through the process of living: a machine gradually coming to bits. We use ourselves up. Death accumulates.

I suppose we could think of it as a glitch. Life is a work in progress, a beta version, inherently fragile and unstable. It's still got a few shortcomings, death chief among them. Somehow, as soon as you boot up, errors begin to accumulate, something hangs up and, eventually, the whole system goes down. But even with that fairly substantial shortcoming, it's still an amazing program, worth the pain and hassle, almost magic. The wonder is not that we die, but that we live at all. The business seems so complicated and improbable, it's amazing that it works. We should, all of us, go around holding our breath from moment to moment, astonished and thankful at our continued existence.

I should add: when I say "magic," I don't mean airy-fairy, Tinkerbell, happy and laughing magic. I mean magic of the awesome, terrifying, lightning and thunder, angry Jehovah kind: scary but powerful. Obviously, life is not all pleasure and delight. Not at all. And surely the time comes, as errors accumulate, when pain must make it all seem hardly worthwhile.

Even so, to have had that glorious chance. To be blessed with that spark of animation, a piece of dirt getting up and walking around, thinking and laughing, falling in love and being aware of all these marvels as they happen. To be given wings to fly, if only for a little while. What a tremendous piece of luck. Who wouldn't die for the chance?

I don't know if such thoughts make this pathos in Thornton Creek

— or my grandfather's suffering — any easier to accept or understand. There's nothing much redeeming about death. What goes up, must come down. Only in dreams do we fly on forever, soaring. But at least I have it in some sort of perspective. The point is not that they have to die — my grandfather, my grandmother, the black bear in the orchard, these magnificent fish — but that they lived at all. And at least for a time they lived gloriously, blessedly, until mortality caught up with them.

And, of course, nothing is wasted. The impact of these lives isn't lost; they ripple onward to some unseen effect. The fish will end up in the forest, many of them, to nourish trees and bears. And somewhere in the gravel, maybe not here, but somewhere, little spheres, bright pink, are quickening, growing toward hatching, the rush to the ocean and out into all the wild world. *Hishuk is ts'awalk; Everything is one.*

MA'MĪQSŪ

(Elder Sibling Moon)

November weather is wet and windy. Days are short and dark; winter has come to stay. The mellow days of Indian summer are a fading memory, the sunny breaks of early spring a faint hope. Over the grey ocean, colder with every sunless day, the wind stirs up great swells and a jagged chop to scourge sand from the beaches. In the mountains, incessant rain runs from forested slopes in sheets and rivulets, gathers into streams, swells the rivers into monstrous muttering torrents.

In the old times, people along the west coast of Vancouver Island reckoned Ma'miqsu was the first moon of the year. Tribes were snugly established in their winter quarters. The salmon harvest was complete and food plentiful. Foul weather made travel difficult and dangerous. With the whole community in one place, it was a time for indoor celebrations, feasting and potlatch. Neighbouring tribes and villages travelled back and forth, attending one another's festivities. Otherwise, winter was a time for preparations, for maintenance of equipment, for carving, weaving and other crafts, for the supernatural rituals that helped to assure success in hunting and whaling.

Even nowadays there is a sense of retreating indoors in November. In an odd way, people relish this coming of winter. After the hectic pace of summer and fall, most are glad of an excuse to spend a quiet day under cover with permission to do nothing much at all. After months of calm, wild weather stirs the blood. The migration has already started as people drift away in ones, twos and threes, searching for winter sunshine in far-away places.

Opitsat

DAY OF THE DEAD: Seeking Y'aq-wii-itq-quu?as
(Those Who Were Here Before)

A moment of truth: six burly men, naked for action, long hair flying, lean forward and dig deep into turbulent water. The canoe surges forward. In the stern, the elder uncle, the mentor, a successful whaler in his own right, orchestrates the hunt, guiding the boat. He watches for just the right moment, when the broad flukes are moving down and away. The young chief stands balanced across the gunwales in the bow, hefting a harpoon, poised to strike down into the swelling smoothness of the whale as it arches up from the ocean below him.

The dark wood, oiled western redcedar, catches and freezes the movement, the emotion, the intensity, the wild heart-pounding energy. It's an enormous carving, almost life-size. The space around is vast and echoing, busy with people coming and going, moving around the periphery, aware, fascinated, but keeping well clear, as always when life and death are played out in public places. And the whole tableau is screened with a fountain of water, from ceiling to floor, so that even indoors it is always raining, raining.

The kayak and I sit secure just outside the surf zone, rising and falling in the gentle swell. An adult bald eagle, white head and tail, sails past on fixed wings, heading low across the beach, patrolling. The rain has eased off and the air is absolutely still, a piece of luck in November when every calm day is a gift. The water is mirror-smooth except for the ripples spreading away from scattered raindrops. I'm wet inside and out but the exertion of paddling has warmed me; I'm not uncomfortable. And rainy-day kayaking does have its charms. There's a good deal of beauty in the moment. The whole scene — the broad bay in front of this beach, the steep, dark-wooded slopes rising from tidewater into the mountain mist — reawakens some of my earliest and most evocative impressions of the west coast. A flock of sea ducks flushes from the ocean and flies off, wings whistling. As always when the weather is wet, the air is thick with the smells of salt water and forest.

Behind the beach is but one small house, where once were many. I look and listen but nobody is home. To judge by appearances it has been a while since anyone lived here; I'm not intruding. The eagle returns and lands somewhere out of sight to the left. The forest up and down behind the beach, the entire length of the cove, is second-growth: young, dense and tangled. The ancient snags and pickle-fork cedars of old-growth forest are missing. Back there, hidden now by vegetation, mounded middens bespeak centuries of occupation.

Evidence of ancient human presence is everywhere in Clayoquot Sound: in the greasy black, shell-laden soil of middens; in the straight sandy canoe runs cleared through sharp intertidal rocks; in piled stone fish traps and weirs. People have been part of this place for many thousands of years, a way of life stretching back clear to the ice ages.

The earliest strata in the dig at Yuquot in Friendly Cove on Nootka Sound have been dated to 4,300 years Before Present (BP). At Bear Cove, near Port Hardy, the earliest layers have been dated to 6,000 BP. And at Namu on the mainland coast of British Columbia, there is evidence of habitation 10,000 years ago.

Every schoolchild knows about the Bering land bridge: with so much of the planet's water locked up in ice-age glaciers, the Pacific Ocean retreated from the shallow floor of the Bering Sea, creating a broad land bridge between Asia and North America, a throughway for groups of nomadic hunters, perhaps the first human beings to arrive on this continent. What isn't so widely appreciated is that other more maritime cultures likely followed the coast, paddling into new territory.

It's not hard to imagine small groups paddling southward almost as soon as the ice had retreated enough to allow passage. Glaciers still poured from the deep valleys, spangling the ocean with pans and bergs of ice. Here and there a rocky point or gravel spit might offer a camping spot. Much the same sort of conditions still exist today, farther north along the coast of Alaska. The land is bleak and inhospitable but the ocean teems with fish, birds and sea mammals, ample resources to sustain a maritime hunting culture. So all through the last 10,000 years — as glaciers melted back, salmon repopulated streams, forests established themselves upon bare rock and gravel — a trickle of humanity ebbed back and forth along this coast.

We know next to nothing about those early inhabitants; they left very little behind. But what little we have — a handful of stone artifacts buried in middens — speaks eloquently. Even in the most ancient layers, the tools are finely wrought. At Yuquot, for example, archaeologists found nicely shaped wedges of stone for working wood, and beautiful little needles with tiny, perfectly shaped holes to carry the threads of sinew. Four thousand years ago, pairs of experienced hands made those tools. Wherever I paddle along this intricate coastline, I cannot help but feel a sense of kinship with the Y'aq-wii-itq-quu?as, the ones who were here before us in this place. What kind of people were they, I wonder? What kind of lives did they lead?

These are not easy questions to answer. Sources are scanty and often unreliable. There's not much to go on.

Recognizable artifacts are rare in all but the most recent strata at archaeological sites. The people who lived here fashioned most of their belongings from wood, shell, antler and bone, perishable materials that decay rapidly in the moist, mild, acidic conditions of these parts;

stone tools are scarce. And it was strictly a spoken-language culture. There are no tangible historical records of those earlier times, no written accounts to tell us about that way of life.

We do have anecdotes and stories of life in the old days, passed down among the Nuu-chah-nulth people. But even in a culture dependent on accurate remembering, a spoken-language culture, stories are shaped and coloured by a story-teller's self-interest and by the social mores of the times, changing as they pass from person to person, generation to generation. And it's been a long time — three or four lifetimes, ten or more generations — since the earliest encounters between local First Nations and explorers of the various European nations. There were enormous social and political changes from the beginning: catastrophic wars between neighbouring tribes, terrible epidemics, a new economy. When today's elders recall their youth or even what they were told of their parents' youth, they are describing a world already far removed from that of their ancestors at the time of contact and before.

There are also formal oral accounts, distinct from casual anecdotes, of genealogy and property rights. In the old times, every young chief, the _haw'iɫ_, would have been taught the history of his family — a genealogy — and the provenance of his property and ceremonial rights, which the Nuu-chah-nulth call _ha-ḥuuɫhi_. In some measure that tradition survived the turbulent times after contact, and continues to the present day. In the old times, such accounts were extremely important; they defined the chief's social position. But these formal reiterations of genealogy and property rights were not histories in the modern sense of the word, they were calculated political statements intended to justify the possessions and authority of aristocratic families. In re-telling the story of his family's past, a chief strove to enhance and strengthen his claim to the rights and privileges on which his own prestige and that of his family depended.

The first written accounts of indigenous culture along the west coast of Vancouver Island come from the journals of early explorers, traders, missionaries and settlers. Our only eyewitness record, they offer the barest glimpse of the local way of life at the time of contact. But early written accounts have their own set of distortions. For a start, indigenous culture had already changed profoundly by the time most visitors arrived. When Gilbert Malcom Sproat arrived in 1850, for example, the First Nations in Barkley Sound had already been ravaged by thirty or forty years of post-contact intertribal warfare. Moreover, as the title of Sproat's book *The Nootka: Scenes and Studies of Savage Life* suggests, early observers had their own set of biases. They absorbed the local situation according to the measure of their own cultures. They could not speak the language fluently and had little or no idea of social contexts. Not surprisingly, they often misinterpreted the cause and nature of events. And most early observers were little more than visitors, not much involved in the Native way of life. They were onlookers, viewing the local culture from safe distances and for short periods of time.

So the knowledge we have comes from no single, authoritative source. It has been pieced together from various sources, comparing each with the others to arrive at some idea of the truth, some vision of the past.

I imagine a long row of houses set on a low terrace behind the beach. The buildings are rectangular, about 10 metres wide and up to 30 metres long. As was the custom on this part of the coast, they are placed with the long side toward the beach — their ridgepoles parallel to the shore — with doorways in the middle of the sidewalls that face the beach. Some of the houses are joined together endwise, to form even larger structures. The roofs are low to the ground and have very little pitch; some seem almost flat. The roofing and siding are enormous cedar planks, some a metre and a half wide, carefully adzed to a finished surface. Roof planks are laid cross-wise, from the ridgepole to the side wall, in two layers overlapping like roof tiles. The siding is laid horizontally, one plank overlapping the next, to shed the rain.

Villages like this would have been loosely organized according to family groups, somewhat analogous to the clans of Scotland. The scholar Philip Drucker[1] called them "lineages" or "local groups." Kinship and inherited rank were the basis of all social relationships. Each lineage was lead by a chief assisted by his close male relatives — his father (the retired chief), grandfather, brothers, uncles, nephews, sons and so on — who, together with their families, represented a form of landed aristocracy with inherited property rights and titles. More distant relatives, with no property of their own, constituted a sort of labouring class, the commoners. The dividing line between minor aristocracy and the commoners was ill-defined. In fact, commoners with particular skills — warriors, shamans, craftsmen — could acquire higher status and minor property rights.

Slaves were at the very bottom of the social structure. They could be born into captivity or taken in war. Slaves were property and, like all other property, were owned by the chiefs. They could be bought, sold, given away or destroyed at will. Chiefs could — and did — kill their slaves to serve a variety of purposes: to demonstrate wealth, to provide a death companion, or to enforce discipline among other slaves. Slaves were compelled to perform menial work, cooking, carrying water, cutting wood, although a slave with special skills was generally allowed to ply his or her trade on behalf of the chief. On the other hand, slaves lived in the same house as their owner, formed part of the extended family and were usually well treated, eating the same food and living about as well as the common people.

The self-identify of each lineage was defined by two things: the family's sense of a common ancestor and their sense of belonging to one particular place in the world, a traditional or ancestral home, which

---

[1]Philip Drucker. *The Northern and Central Nootkan Tribes.* Bulletin 144, Smithsonian Institution, Bureau of American Ethnology, 1951. A major work of scholarship based on field work conducted in 1935–36. Drucker collected records from his own informants and compared that information with the other sources to prepare a picture of First Nations life in the period 1870–1900, with references to earlier times.

might be a certain beach, a cove or a fishing place on a river. Family prestige was the key to power and wealth, and vice versa. Each chief strove to uphold or enhance the family name. Several different clans or lineages might ally themselves to create what Drucker calls a tribe. Among these different lineages there was an established hierarchy, according to the relative prestige of each house or clan. The tribe often gathered in winter at the traditional place of the leading lineage, and the whole group would adopt the name of that lineage. In more recent times, tribes often merged into even larger "supergroups" or "confederacies," again taking the name of the leading lineage. This multiple employment of certain names leads to some confusion. A given name, Moachat, for example, might refer to a single lineage or clan. Or it might refer to the tribe dominated by that lineage. Or it might refer to the supergroup or confederacy of which that lineage was the leading member.

Eugene Arima[2] names some 34 independent political units from the first half of the 19th century: the Chickliset (tribe), the Kyuquot (confederacy), the Ehattisaht (confederacy), the Nutchatlet (confederacy), the Mowachat (confederacy), seven independent local groups along Muchalat Arm and the Gold River valley, four or five independent local groups in Hesquiat Harbour, the Manhousat local group in Sydney Inlet, the Otsosat (tribe), the Ahousaht (tribe), the Kelsemat (tribe), the Tlaoquiaht (tribe), the Ucluelet (tribe), the Toquaht (tribe), the Uchucklesaht (tribe), the Sheshaht (tribe), the Opetchesaht (tribe), the Ohiaht (tribe), the Nitinaht (tribe), the Clo-oose (tribe), the Carmanah (local group), the Pacheenaht (tribe), the Makah (tribe), the Ozette (tribe). Oral history suggests that there were many more politically distinct units — perhaps twice as many — before the arrival of European explorers in the 1770s.

---

[2]Eugene Y. Arima. *The West Coast People: The Nootka of Vancouver Island and Cape Flattery*. British Columbia Provincial Museum Special Publication No. 6, Victoria, B.C., 1983.

Each house on this beach would have belonged to a separate lineage or one of the junior branches. Each was the property of a chief, who lived more or less permanently in the house with his close relatives. But commoners with blood ties to several houses were free to come and go. If they felt mistreated they could seek a more congenial home: a house where the people were friendly, the food was good and the work not too daunting.

This freedom among commoners to shift their allegiance to other houses created a balance between aristocracy and labour that profoundly shaped everyday life in the community. The chiefly class owned everything: property, resources, houses, names and songs. But in a labour-intensive society, ownership was useless without a supply of labour to harvest the resources, to create the goods that wealth could buy, to display the songs and privileges. Even though chiefs were theoretically all-powerful, their power was curbed by the need to stay on good terms with their tenants, who in turn depended on the aristocracy and their property for food, shelter and status. From earliest childhood, chiefs were taught to get along with their people, while the children of commoners were taught to look after the children of chiefs, to respect their authority and to work at enhancing the prestige of their chief.

I see smoke. Wisps of smoke rise through gaps between the roof planks of houses along the beach, suggesting life within. Old men and women are gathering on plank platforms set up in front of the houses to get some fresh air, to enjoy this lull in the storm. They squat, talking, wrapped in cedar-bark blankets, their backs to the house walls. Two men carry a log of wet firewood up from the beach. Children race in and out around the houses, their shrill voices echoing across the water. A group of women tends to a cooking pit, steaming the last of the wild clover roots or a mess of clams. Among them, a teenage girl wears the elaborate dentalium hair decorations of puberty — she's not really supposed to be outside, but at least she's well chaperoned. Down at the water's edge, two fishermen unload a canoe. Another heads for the houses with three or four big cod on a withe. The boys are roughhousing down at the other end of the beach. Two of them have their hands in one another's hair, trying to wrestle each other to the ground.

Suddenly everyone looks in my direction.

Two canoes are coming into view, just out past the rocks. They are lashed together side by side. Two or three planks laid across the gunwales make a platform. On the platform, two dancers are performing: thunderbird and lightning serpent. The paddlers are beating time and singing. Another canoe follows the first two coming around the point. A young man stands in the bow holding a whaling harpoon. He dances, going through the motions of striking a whale. The canoe is bouncing with rhythm as the paddlers drive it forward. Yet another pair of rafted canoes comes into view with two more dancers: killer whale and wolf this time.

This is surely a marriage party. These people have come to propose a marriage. Those on the beach and in front of the houses are laughing now and calling out. I look everywhere up and down the row of houses, but the young woman with the elaborate hair decorations has vanished.

The eagle calls from its perch, a surprisingly feeble twitter for such a regal bird. Perhaps its mate is somewhere nearby, out of view. Time to get a move on. November days are short and I have some ground to cover. Fortunately, the trail is just around the next headland. I haul the kayak onto the drift logs beyond the reach of high tide and tether the bow to a tree for good measure. I squirm out of my wetsuit and change into hiking gear, including heavy PVC rain suit and caulked boots.

I start briskly, enjoying the walk very much, looking forward to my destination, a place I've read about but never visited. After confinement in a cramped kayak, it's a relief to be stepping out, stretching my legs, breathing it all in. It has started to rain again, but no matter. The forest is even more beautiful after a bit of rain. Every twig and leaf is bejewelled and gleaming with water. The colours are vivid. The air is fresh and full with the scent of damp earth and wet wood. The world is rinsed clean, freshly bathed, exquisite.

Soon I'm perspiring heavily. The going is rough. The path is muddy and treacherous with deep greasy wallows and slippery roots. I have little choice but to follow it; the forest is absolutely impassable, dense and tangled. Under a leaden sky, I slog down the bottom of a green ditch of cedar, hemlock and salal, lurching like a drunkard through the mud holes, scrambling over tangled roots and windfall. More than once the sharp caulks on the boots save me from a nasty fall. Soon I'm wringing wet, inside and out. The trip seems endless. I find myself wishing that I'd stayed at home, warm and dry.

That's exactly what the west coast people did at this time of year. November weather is generally too miserable for outdoor work. So this was the time of year for indoor entertainments, feasts and potlatches.

Simple feasts were casual affairs, thrown at short notice, often to celebrate some minor occasion. When a little boy killed his first game or a girl helped to gather clams or cedar bark for the first time, the event might be marked by a feast at which the child would receive praise and recognition. When a chief received a good harvest from one of his properties, a berry patch, a salmon stream or a drift-whale beach, he would invite the people to a feast. Common people could do this also. If a man went trolling and caught some salmon, he would invite a group to eat with him.

Entertaining was one of life's great pleasures. Then as now, giving a feast raised the host's reputation among his friends. For a chief, feasting served an additional purpose, more subtle and serious. In theory, the chiefs owned everything. But the ownership of resources was of no use without willing help to bring in the harvest. So the chief would ask his people for their help, invite them to make the harvest. When they returned with the produce, he would reward them with an occasion: food and entertainment. He could also use the festivity to assert his ownership of the resource. During the feast, or afterwards, the chief would thank the people for gathering the harvest from his _ha-huułhi_. He might recount the way in which he'd gained the rights to that particular property, usually as part of an inheritance or dowry. He might speak of the potlatch in which his father had passed the property to him. He might tell them that they were welcome to help themselves for the rest of the season; he would give his permission. But he would ask them to please remember that this was his property and to take good care of it for him. It would have been all easygoing and lighthearted. But by eating the food and taking the presents that sometimes came at the end of the feast, the guests were tacitly acknowledging that chief's claim to his property.

Potlatches served a similar purpose, but on a much more formal level. Potlatches required more planning and preparation, and were reserved for especially significant occasions: birth, puberty, marriage, death. When a chief passed rank and property to his son, for example, he would mark the occasion with a potlatch. When a chief potlatched, he was performing for a much wider audience than at an ordinary feast. He was addressing not just his own people but other lineages and tribes. Specifically, he was entertaining chiefs from other clans and tribes, fellow aristocrats, and asking them to recognize and validate his ownership of certain properties and privileges, and thus his right to bequeath those properties to his heirs. In keeping with the greater dignity of the guests, more food and more elaborate entertainments were required to keep everybody in a friendly and agreeable mood.

In a proper potlatch, the guests would be seated with pomp and ceremony, the ushers paying strict attention to seniority and status. Guests were fed on a lavish scale, far more than they could eat; in fact, they were expected to take the leftovers with them. At home they would use these leftovers to give a feast at which the events of the potlatch would surely be recounted, helping to spread the chief's claims even further. During or after the feast, the host would display his privileges in dramatic fashion: dances, songs, names, masks. Some of these would be valuable in their own right. Others might represent or be associated with less portable property. A mask, say, could represent an island that the chief owned and was now bequeathing to his young son. A song might be attached to a *topati*, a test of strength or skill a suitor had to pass if he was to marry the chief's daughter. When the privileges had been displayed, the host would thank his guests for witnessing his ownership. Then, to seal the bargain, he would give away property, assisted by his speaker, the tally-keepers and young men who carried gifts to the chiefs.

Of course, the attractions of potlatching went far beyond the formal display of property rights. A potlatch was an opportunity to have some fun in the dreary days of winter. It was a breath of the dramatic, of mystery, wonder, theatre. Potlatch satisfied a love of ceremony and ritual; it bound the people together. The giving of gifts back and forth made for good feelings among the clans and tribes. It was a way of distributing wealth from haves to have-nots. And it was a way for the chief to enhance his reputation. By giving a good, generous potlatch, the chief could make a name for himself, his immediate family and his lineage. Potlatching on the west coast of Vancouver Island seems to have had little of the competitive flavour of potlatches elsewhere along the coast of what is now British Columbia, where rival chiefs sought to embarrass and outdo one another in the ostentatious display of wealth. By all accounts, the whole idea here was to render guests happy and content, not to embarrass or shame them.

When the distribution was complete, the highest-ranking chief among the guests would act as a spokesman, thanking the host on behalf of all present. He or his speaker would say, essentially, that they all knew the chief's words to be true, that the privileges he had shown were rightfully his to bequeath to his heirs. The spokesman might add that the chief was a great man and that, potlatching on such a lavish scale, he had done much to uphold his family's name and the honour of his ancestors.

The sky is brightening little by little. Astonishing. By the time I reach the edge of the forest and climb out of the brush onto the drift logs behind the beach, I can see patches of blue. I shed my rain gear, hoping that sodden clothing might dry slightly if I get some air to it. I step down onto the sand, happy to be in the open, clear of the trees at last.

And the view is spectacular: a panorama across the low rolling country behind the beach, sunlit now, with the dark mountains of Clayoquot Sound in the background, peaks still shrouded in heavy cloud. The bay is fairly well sheltered by a cluster of rocky reefs offshore, but even so, an impressive wintertime swell is pounding in. A short scramble beyond the end of the beach takes me to the rocky point at the head of the bay, my objective for the day. Behind the rocks is a dense tangle of second growth, mostly salal, salmonberry and a few young conifers. I see no obvious sign of human habitation, but this does not surprise me. Nature reclaims her own very quickly in this climate. Besides, I've no idea where this village was actually located; I wouldn't know where to look for any remains.

But it's easy to see why the people would have chosen this spot for their home. With the little islands and reefs as a sort of natural breakwater, the place is surprisingly sheltered. Skilled paddlers could land here in almost any conditions. And think of the resources. Above the tide line, all the riches of the forest close at hand. The beach behind the bay, the little coves flanking it and all the rocky shores in between, a broad buffet of intertidal seafood. The bay itself is prime whale habitat. The halibut grounds are not too far away. And the rocky reefs just offshore provide excellent salmon fishing, some of the best in Clayoquot Sound even today. Only one thing is missing, a fatal flaw: no spawning stream for salmon.

As always in such places, I wonder again what these people would have been like, just as ordinary folks, friends and neighbours. What kind of life did they lead out here?

For one thing, they would have been very much in tune with the natural world around them, alert to every little sign, like modern-day fishermen and farmers, like people the world over who must daily make their way in the out-of-doors at the risk of their lives, gathering food, materials, supplies. Making a living. It must have been a rough life. They would have been rough people — hardened, strong and competent — who drew their living from a harsh, unforgiving environment using meagre tools of stone, bone, shell and wood. I remind myself that they were not playing games; they were not out here for the fun of it. This was their life, their job, their living. They were expert at what they did, highly practised. Think of the most able man you know and the most competent woman. There would have been many people here like that: same talents, different circumstances. But it must have taken a lot out of them.

In their home life, by all accounts, they were remarkably good-natured. Mildness of temper, lightheartedness and generosity were the essence of good character. Philip Drucker's informants would tell him: *So and so was a good man. He never quarrelled with anyone, and was always laughing and joking. He would always help people, and invite them to eat with him. Everyone liked him.* Children were taught to be kind and helpful to others: *If you see a man pulling a canoe up on the beach, go help him without his having to ask you.* They were taught not to quarrel with others in the community: *If somebody says something mean to you, just walk away. A real chief doesn't squabble.* The children of a chief were taught to take good care of the common people, providing food and giving feasts. Children of lower rank were taught to play carefully with the chief's children, to help them and never to quarrel. The people worried a good deal about what their neighbours and relatives thought of them. They valued their good name and reputation. And they very often suppressed their own ambitions and hopes in favour of the common good. They loved ceremony, feasting and a good laugh.

On the other hand, in time of war — and war was commonplace — these same people could be terribly cruel and ruthless: the absolute opposite of their ordinary selves. Very often, a war began as the personal project of one man or a small group within the community, who took it upon themselves to rouse their neighbours against an enemy. There was nothing symbolic about warfare on this coast, no counting coup. The goal was almost always to annihilate an enemy. Men, women and children were killed, though women and children might also be taken as slaves. The special horror of these conflicts stems from the strange and terrible intimacy of the bloodshed. Very often enemies lived in sight of each other; villages on either side of a channel might be locked in a bloody struggle for survival. Warfare was close and personal. Enemies knew each other socially. Often people on one side of the conflict had close relatives on the other side.

Conflict was driven, on the surface at least, by passionate emotions: a fear of attack, a rage for revenge. But there were often other issues. In this place, for example, it came down to the lack of a salmon stream. In summer everything would be fine. The sea was calm. Men could fish or hunt sea mammals. But without a salmon stream, winters would have been hard. The people would be forced to get by on dried halibut, then mussels, clams, whatever else they could find, until the herring came in. The neighbouring tribe — on any clear day you would have been able to see the smoke from their fires over there across the channel — controlled several salmon streams. The people who lived in this place tried to gain access through marriage, tried to get one of those streams as part of the bride-price. When that failed, they resorted to war on the pretext of an insult supposedly received from the other tribe. They succeeded in annihilating their neighbours and took that territory with its streams, leaving this place and its ghosts to begin a slow slide toward obscurity.

I explore the point and the nearby beaches until it's time to head back. I don't plan to linger. The sun is still shining brightly but I'd like to get home while the weather holds, preferably in daylight. It's been a good trip, a safe trip, and I want to keep it that way. It's a successful trip too: I have a much better sense of what life might have been like for the people who lived here, my goal for the day accomplished. Even so, I can't help feeling a faint disappointment. I was hoping, against all reason, for some clear sign of human habitation, a direct link to those people who were here before. I know that I'm chasing a will-o'-the-wisp, but I can't help it. It would be nice to have something tangible, something I could touch with my hands.

I'm just starting back, making my way cautiously along the top of a drift log, one of an immense pile at the head of the beach, when I notice a game trail going off into the bush, exceptionally well worn, with just the suggestion of an opening back there. I'm a sucker for these secret hidden places. Perhaps there's a little marsh, a lagoon or a sandy dune, a nice camping spot for some future trip. Probably it's nothing at all. I've been fooled many times, seduced by likely looking trails to nowhere. I'm anxious to be on my way; I'm already thinking about the kayak. But there is something about this trail that intrigues me, something inviting. I check my watch. *Five minutes*, I tell myself, as I step into the bush.

What I find, when I finally pull free from the salal, amazes me. A grassy clearing scattered with clumps of spruce stretches back a hundred metres or more, park-like, in a series of low rolling terraces, five to ten metres high. There are one, two, three of these, rising back into the forest, one behind the other. Sunshine illuminates the grass. The whole meadow is a brilliant green. A quiet, peaceful place.

This is how it must feel to stumble across a lost city or a forgotten temple. On the west coast of Vancouver Island, amid the tangle and riot of rainforest vegetation, a large grassy clearing is an extraordinary anomaly. It's like coming upon a patch of jungle in the middle of the desert or a tropical beach in the Arctic: a bit eerie, mystical, certainly haunted. I feel more than a hint of the forbidden world here.

The five-minute limit forgotten, I wander into the open for a better look. Where the meadow finally gives way to forest, the terraces continue, disappearing under the second growth. It's an impressive site, though far from undisturbed, as I soon discover. The forest all around was logged some time ago. There are deep wheel ruts through the clearing and wide trenches where roadways were cut through the terraces.

These terraces are no work of nature; they're huge middens. Here, the debris and refuse of this village accumulated over hundreds or perhaps thousands of years. In places, the grass and moss have fallen away from the steep banks of the road-trenches to reveal the typical ink-black soil, greasy with organic matter and shot through with little bits of seashell, so characteristic of west coast middens. These particular middens have become so large they're interfering with the natural drainage. Small patches of marsh have developed in the hollows between them.

When this place was inhabited, houses stood in rows along the top of each midden-terrace. I look but cannot see any timbers remaining. This is no surprise. Untreated wood does not last long in this climate. But at regular intervals along the top of each midden are clumps of young Sitka spruce. As I approach more closely, I realize that each little cluster is laid out according to a distinct rectangular pattern, very similar from clump to clump; these trees are nursing upon the remains of corner posts and ridgepoles.

It's not hard to imagine the houses as they were — the great posts and beams (huge logs, some freshly adzed, the smell pungent) and the beautiful planks of redcedar — along with the movement and life of the village. In summertime, one might see fishermen unloading halibut from their canoes down on the beach; a group of women going off with burden baskets on their backs to gather berries or eel-grass roots; a canoe being paddled through the reefs offshore. At this time of year, on a day like today, there would be people outside enjoying this break in the weather. Women might even bring their weaving or basketry outdoors. At the far end of the beach, there might be a couple of men using the sunny afternoon to put final touches on a new canoe, pitching the seams or scorching the hull to smooth it and remove the slivers.

I sit on a large branch that grows low along the ground from one of the corner-post spruces, sticky with pitch but comfortable. The view down the meadow is especially lovely. A small airborne column of insects dances in the sunlight nearby. Where do they come from at this time of year, I wonder? They must tough it out all through the rainy days, waiting and hoping for the sun to shine, just like the rest of us.

Typically, a historic site focuses on a brief era, sometimes just a single event, a crisis, an extraordinary but passing phenomenon. But in this place we get a very different perspective. Here is an extraordinary accumulation of ordinary life, daily events repeated thousands upon thousands of times, over thousands and thousands of years, gradually aggregating like layers of earth, like the rings of a tree. That is what speaks to us. Even now, I can feel the weight of all that humanity, a great multitude if they could but gather in one time and place. The meadow is practically humming with human presence, slightly muted, as if felt from a great distance. It is not alarming, not frightening. I have a sense, you know, of being *welcome*. Perhaps they're pleased that I've come to visit; I don't suppose anyone else has dropped by for a good long time. It's a beautiful spot. I don't want to leave. It occurs to me that I have, in a minute way, become part of this community, part of humanity's relationship with this place, by simply finding my way here today. The thought pleases me very much.

Who were these people: hundreds of generations, occupying this place over thousands of years? Nobody knows. Nobody can know.

There is no record. The only message they've left us, the sign and symbol of their passing, is in the accumulated earth of these great middens. But I feel a powerful sense of kinship nevertheless. Perhaps that is what gives this place such significance. A local site like this echoes a specific local aboriginal culture, certainly, but it also speaks more broadly of simple human endurance under difficult circumstances. In that sense, all humanity has a stake in what happened here. This is part of the story of our species, testament to the everyday heroism, strength, tenacity of which human beings are capable, holding this storm-swept point at the edge of the north Pacific for millennia, a source of pride and admiration for us all.

Places like this also echo a tradition, a way of life, that in its finest elements is not peculiar to one First Nation or another but is a reflection of the country itself — a quiet, simple life between ocean and forest on the wild west coast, a life lived in a fair degree of harmony with the natural world. It is a tradition that many still aspire to, among First Nations and Mamułni alike. In that sense, the people who were here before — the Y'aq-wii-itq-quu?as — are forebears, ancestors, to us all: even to those who may have migrated rather more recently than the ice ages.

Too late, as I sit here, it occurs to me that I'm probably trespassing, not against those who were here before but against the present-day owners. A place like this has got to be a reserve, surely, private property. I respect that ownership and regret the intrusion. Even so, I'm not sorry to have seen it. Maybe I even have some sort of modest right. Somebody else owns this property, no argument. But — with all due respect to a people who have every reason to suspect the acquisitiveness of newcomers — I'm not persuaded that they own the past it represents. That belongs to every one of us as we take our place in the story.

Along the farthest boundary of the meadow is a line of very large stumps, perhaps a memento of the drive for Sitka spruce wood during World War II. Wouldn't it have been something to see them towering over the village? I'm fifty years too late. But then I discover — down at the very end of the line — a living giant. For a moment, I wonder how it escaped. Then I look up and discover that the main spar is broken off not too far above the ground, casualty to some ancient storm. Deformity saved this tree's life; it wasn't worth taking. Around the snag, the living branches have turned upward to become trunks in their own right.

The main trunk — I reach out and touch it — has a diameter of at least three metres. It's far larger than anything growing down on the meadow; must be four or five hundred years old, a witness to daily events all through that time, my tangible link with the past. I listen respectfully, just in case this tree has anything particular to tell me on this day. Nothing comes through, but no matter. For me, it is enough to be here. I'm content with the day, content with myself. All is quiet and peaceful in the sunshine. The sound of surf is muted. I stand next to the tree and look down across the village, thinking about what this tree has seen here, watching patiently day after day, year after year, century after century. Above all, I feel the comfort in this place. More than any other experience in this year of rediscovery, this abandoned village embodies the essence of what I was chasing when I came to the west coast of Vancouver Island all those years ago. Perhaps I have come home at last.

December

QAⱢATIK
(Younger Sibling Moon)

DWinter deepens in December, the darkest month of the year. Storms pound the coast. Everything is wet, constantly; the ground is saturated with moisture. On the mountain slopes above the inlet, the snow line settles lower and lower, driving forest birds down to the water's edge. Out on the inlet, ducks dabble and dive.

In the old times, the people had no calendar in our modern sense of the word. They divided seasons according to lunar months, each with its own traditional pursuits. This thirteen-month cycle was also divided into two halves. In one half, the days grew longer. The sun rose farther and farther northward, until at last it rose in the same place for four days, so it was said. Then it would start back in the other direction. The days grew shorter. Every morning, the sun rose a little farther south along the horizon until finally it rose in the same place for four days and the pattern repeated itself. In Qaɬatik, festivities continued, feasts, potlatches and other celebrations. But soon after winter solstice the clans were on the move again, heading for winter village sites near the quiet coves where herring and the salmon that prey on herring would be arriving over the coming weeks.

In modern times, of course, December is also a festive season. I stepped onto the beach one morning to find a big coast-guard buoytender anchored out in the channel, making a seasonal visit to the light station. The vessel had a Christmas tree fixed to its radio mast and I could imagine Santa Claus — dressed in coast-guard colours, naturally — making his rounds. Some people stay home. Others go out. Christians celebrate Christmas. Neo-pagans celebrate solstice. It all works out. With luck, everyone has a pleasant time. In the words of the old carol: "The country guise is then to devise some gambols of Christmas play; Whereat the young men do the best that they can, to drive the cold winter away."

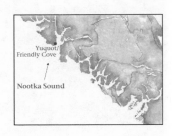

Yuquot/
Friendly Cove

Nootka Sound

A MEASURE OF HISTORY:
In Search of the Recent Past

Somebody composed a song right away, while they were
still on the ocean. The song says: "I got my walls of a
house floating on the water." Since those ships were
found floating on open Pacific, we started calling the
white people Mamaɬni, regardless of what nationality,
even the white people that have never been on the water.
— Peter Webster in *Nu•tka•: Captain Cook and the Spanish*
Explorers on the Coast, 1978

The flight is uneventful. The weather is good: a mid-winter break. We fly on, engine droning, the aircraft taking us farther and farther out to sea. Clayoquot Sound falls behind. It's Nootka Sound now off the starboard wing. Far below, a northwest swell crawls across the wrinkled immensity, heading toward its moment of truth with that distant shore. I'm still looking for ghosts — a different tribe this time.

Historian and geographer Brian White, author of a concise analysis of outer coast Vancouver Island history,[3] uses an intriguing term: the relict landscape. The relict landscape comprises all the abandoned and decaying physical evidence of previous settlement patterns, economic and spiritual activities, the whole range of human undertakings. The historical geographer begins with that physical

---

[3]Brian White. *Human History*. Special Studies, Pacific Rim National Park, 1974.

evidence, the artifacts, and works back in time, teasing out history, separating the overlapping enterprises, layer upon layer compressed together in the relict landscape. The point is that those artifacts are not simply the ragged remains of defunct physical structures, but the tangible evidence of human endeavour.

It's an interesting idea. Consider a set of footprints in the sand. We haven't seen the walker — small feet, bare toes, high arches — going up the beach; we've missed the actual event. But we can see the lingering traces and from those we can reconstruct something of what happened, embellished by imagination perhaps. The relict landscape becomes a sort of cumulative, fragmentary after-image, a visible but fading memory superimposed on the present, like a photograph unmaking itself. We can use that fragmentary after-image to further our knowledge of past events. We can also use it to anchor our knowledge of past events in the physical landscape — to give history a home.

I confess to a certain bias: I'm not much enamoured, generally speaking, with those fragments of the countryside that qualify as relict landscape — the fragments that human beings have had their way with. To my eyes they generally have a desolate, neglected, diminished appearance: bits of wasteland with not much to recommend them. I reserve my real enthusiasm for wilderness. But modified landscapes are undeniably a part of Clayoquot Sound and they deserve some attention. Occasionally their power and resonance surprise us; think of the old village site near the point in November.

So that's what we're up to this month: flying north, looking for bits of the relict landscape, hoping to raise the dead.

Forty nautical miles (75 kilometres) offshore and almost due west of the entrance to Nootka Sound, we bank and spiral downward, looking for one particular point on the surface of the ocean. The relict landscape is faint here. The event in question left no physical trace at all, barring a passage of written description:

SUNDAY 29th [of March, 1778]. At length at 9 o'clock in the Morning of the 29th as we were standing to the NE we again saw the land, which at Noon when our Latitude was 49°29'30" N, longit. 232°29' East, extended from NWBN to ESE the nearest part about 6 leagues dist^t. The Country had a very different appearence to what we had before seen, it was full of high Mountains whose summits were covered with snow; but the Vallies between them and the land on the sea Coast, high as well as low, was cloathed with wood. The SE extreme of the land formed a low point off which are many breakers, occasioned by sunken rocks, on this account it was called Point breakers. *It lies in the latitude of 49°15' N, longit. 233..20" [E] and the other extreme in about the latitude of 50 & longit. 232°00 E: this last I named* [Woody Point], *it projects pretty much out to the SW and is high land. Between these two points the shore forms a wide bay, which I calld* Hope bay, *in which, from the appearance of the land, we hoped to find a good harbour and the event proved we were not misstaken.*

— Captain James Cook, journal entry for March 29, 1778

But it's enough, isn't it? With a little enhancement we can see what's going on. Captain Cook's *Point Breakers* corresponds with modern-day Estevan Point. *Woody Point* is the modern-day Brooks Peninsula and Cape Cook. The squadron, the ships *Discovery* and *Resolution,* has been at sea for almost two months since leaving the Sandwich Islands (now Hawaii) on February 2. For the last three weeks, Cook and his crews have been tacking back and forth off the west coast of North America, battered by unrelenting March gales, not daring to close with the unknown lee shore in such stormy weather. They are very short of water and the *Discovery*, poorly refitted at the beginning of the voyage, is desperately in need of repairs. With this break in the weather, Cook is determined to make landfall.

The plane levels out above the ocean, headed east. We follow *Discovery* and *Resolution* toward the distant, mountainous coast.

This is Cook's third long exploratory voyage to the Pacific. These are not commercial trips or imperial voyages made with the intent of claiming new territories, but genuine research expeditions — though they do undeniably have a strategic dimension to them. Cook's first voyage brought him to Tahiti, where his crews observed the transit of Venus across the face of the Sun for the Royal Society, gathering data that astronomers needed to calculate the distance of the Sun from the Earth. He and his crews observe and chart. They test the Admiralty's new chronometer. They make the acquaintance of local peoples and observe local customs.

Their present task is to explore the west coast of North America with a view to finding the western entrance to the Northwest Passage between Pacific and Atlantic oceans.

Cook's three voyages — along with comparable expeditions by Don Alejandro Malaspina, Jean François Galaup, the Comte de la Pérouse and others — are a product of the current climate of scientific interest in Europe; it's the Age of Enlightenment. There is a growing sense that better understanding and knowledge could release the world's unexploited resources for the benefit of all. These are broad-minded efforts with relatively free exchange of information gathered; no small thing considering the political and commercial rivalries of the day.

But there are major risks. Both Cook and his second-in-command, Captain Charles Clerke of the *Resolution,* will be dead within a year and a half, Cook of stab wounds and Clerke of tuberculosis. La Pérouse and his entire expedition will be lost in a storm north of the New Hebrides. Malaspina will make it home alive, only to be imprisoned by his king. Dangerous commodity, knowledge, to seek and to hold.

Cook and his crews proceed cautiously toward the coast, taking regular soundings, feeling their way in. Closer to shore, it becomes apparent that there are at least a couple of inlets to choose from. Cook makes for the nearest; so do we.

Nootka Sound is a narrower, more confined inlet than either Clayoquot or Barkley sound, but it has the advantage of a relatively deep, unencumbered entrance, which must have endeared it to sailors working these uncharted waters. The mouth of the inlet is a passage between Yuquot Point on the west, part of Nootka Island, and Burdwood Point on the east, at the northern limit of the Hesquiat Peninsula. The main body of the sound, surrounded by mountainous country, is occupied by a largish island, Bligh Island. Several smaller inlets off the main sound occupy deep, narrow valleys running back into Vancouver Island.

Discovery and *Resolution* lie becalmed for a time in the open ocean off Nootka Island. Then a breeze springs up, blowing fresh from the northwest, and carries them through the narrow mouth of the sound. In sheltered water at last, the *Discovery* anchors in 85 fathoms and "so near the shore as to reach it with a hawser."

We fly up the middle of the channel. Burdwood Point is to the right. Yuquot Point and the Nootka light station are to the left. Friendly Cove and the village site of Yuquot — not much there now — are behind the point. To the north, the looming slopes of Nootka Island rise from the middle distance.

Cook and his crews soon discover that they are not alone:

> We no sooner drew near the inlet than we found the coast
> to be inhabited and the people came off to the Ships in
> Canoes without shewing the least mark of fear or distrust.
> We had at one time thirty two Canoes filled with people
> about us, and a groupe of ten or a dozen remained along
> side the Resolution most part of the night. They seemed
> to be a mild inoffensive people, shewed great readiness to
> part with any thing they had and took whatever was offered
> them in exchange, but were more desireous of iron than
> anything else, the use of which they very well knew and
> had several tools and instruments that were made of it.

As trade progresses, brass becomes a favourite item of exchange, so much so that by the time the squadron leaves four weeks later:

> ... hardly a bit of brass was left in the Ship, except what was in the necessary instruments. Whole Suits of cloaths were striped of every button, Bureaus &c^a of their furniture and Copper kettle[s], Tin canesters, Candle sticks, &c^a all went to wreck.

The Natives prove to be:

> ... very keen traders getting as much as they could for every thing they had; always asking for more give them what you would, neither would they depend on their own judgement but Ask the opinion of all their friends & companions: handing what you offered them round for every one to see it, & then return it to us if they did not like it.
>
> — J. T. Bayly, in *Captain James Cook,*
> *A Voyage to the Pacific Ocean,* 1784

In return, the crews receive a range of items including "the Skins of various animals, such as Bears, Wolfs, Foxes, Dear, Rackoons, Polecats, Martins and in particular the Sea Beaver." Apparently Cook's men collect the skins primarily as souvenirs of their visit to this wild place or perhaps with some idea of procuring a little extra warmth for the Arctic waters ahead. It isn't until much later and by pure chance that they learn the true value of "Sea Beaver," now better known as sea otter.

These are Moachat from the village of Yuquot, later called Friendly Cove, just inside the mouth of the sound. Evidently some confusion arises between Mamaⱡni and Moachat over place names. Initially, Cook calls his harbour King Georges Sound. But somehow he comes to understand that the local name for the place is Nootka and changes the charts accordingly. But *nu•tka* is not the name of a place. It is, rather, a form of instruction or direction, meaning *go around* — around the point into the harbour or around the island. It's easy to imagine a situation where such confusion might arise.

In any event, Nootka becomes the name of the place and by extension the name of the people who live there — the Nootka Indians — and ultimately the name for Native people living all along the west coast of Vancouver Island, until in modern times they chose to call themselves the Nuu-chah-nulth people. Present-day commentators like to make fun of Cook's error, but the man surely deserves some credit for trying to discover and use a local name. Maybe he should have stuck with King Georges Sound; he wouldn't have to take any flak from posterity. Certainly those who followed felt no such scruples: the landscape is littered with imported names.

Cook's crews were not the first Mamałni that the Moachat had encountered. Don Juan Pérez stopped briefly outside Nootka Sound in August 8, 1774, less than four years before Cook, on his way back from a voyage that had taken him to 55° N. *He* called the sound Boca de San Lorenzo. Pérez and his crew are credited with having been the first Mamałni to make landfall along the coast of what is now British Columbia: Langara Island off the northwest corner of the Queen Charlotte Islands, July 1774. Contact between Pérez and the Moachat was brief. None of the Mamałni came ashore and a growing storm cut short the visit. Rather than risk being trapped against the lee shore, Pérez weighed anchor and headed for deeper water.

These first encounters between Mamałni and Moachat seem to have been remarkably civil. There were frictions and a few incidents, but on the whole it was a promising start. The Moachat were hospitable and friendly. The European explorers were experienced observers of different cultures, inclined to be open-minded and tolerant. The two peoples got along fairly well.

The aircraft passes low in front of Cook's anchorage at the south end of Bligh Island. No trace of the relict landscape there. It looks like a wilderness cove, absolutely untouched, though it matches the drawings made by Cook's illustrators, particularly the famous drawing of the *Discovery*, dismasted, in the midst of her refit. We continue around the island, a counter-clockwise tour of Nootka Sound, past the mouths of the various inlets and so back to Friendly Cove. I wish I could say that the countryside looks much as it did in Cook's day, but it isn't so. There are some enormous clearcuts back there and a little snow makes them stand out all the more distinctly. That, too, is part of the relict landscape, sign and symbol of my tribe.

There is also plenty of relict landscape at Friendly Cove, but it's blurred and confused by overlapping layers, era upon era. Here too I can see traces of scenes that Cook's illustrators caught in their drawings: the village site rising step by step in midden-terraces, the snug harbour behind a sandy tombolo that runs out to a rocky headland. I can even see the little promontory where the Spanish garrison would later place their battery of cannon.

In November of 1779, a year and a half after leaving Nootka Sound, Cook's squadron reached Macao. There they discovered that the sea otter pelts they'd been using for hard-weather clothing and bedding were worth a fortune. One pelt, traded originally for a broken brass buckle, fetched three hundred Spanish dollars, an enormous sum. Cook's officers and crew had tales to tell on their return to England. But it was the publication of Corporal John Ledyard's *Journal* in Hartford, Connecticut, in 1783, and the official account of Cook's voyage, published in 1784, that really let the cat out of the bag. The first trading expedition under Captain James Hanna, a vessel with the appropriate name of *Sea Otter*, arrived less than a year later. British ships, some of them commanded by Cook's former officers, had the field pretty much to themselves for the first few seasons. But the first American ships arrived in 1788 and soon dominated the trade.

Trouble is that the maritime traders were a different breed of men, not always as tolerant or open-minded in their dealings with Native peoples as Cook or Pérez. The increasingly fierce and competitive fur trade brought out the worst in all parties. There were ugly scenes and loss of life on both sides.

In 1789, early in May, the Spanish — who claimed possession of the Pacific coast of North America under the 1493 Papal Bull of Alexander VI and now saw their control slipping away — returned to Nootka under Don Estéban Martínez. They took possession of the port, established a stronghold on the site of the Native village and attempted to assert Spanish authority over the traders. The quick-tempered Martínez was the wrong man for a delicate job. His arrest of the equally bad-tempered British trader James Colnett and the seizure of the latter's ships quickly ballooned into an international incident that carried Britain and Spain to the brink of war when Colnett's employer, John Meares, demanded restitution. Nootka Sound, as obscure as the far side of the Moon a bare decade earlier, was suddenly a topic of conversation in the salons of Europe.

In the immediate aftermath of the arrest, the Moachat Chief Callicum, who had taken Colnett's part, was killed as the result of a misunderstanding. This was particularly grievous considering that the Spanish, especially Martínez's successors, Don Francisco Eliza, Don Pedro Alberni, José Mariano Moziño and Don Juan Francisco de la Bodega y Quadra, seemed to be decent men who got along well with the Moachat and their famous chief, Maquinna. Like their British counterparts, the Spanish officers were devoted to research and exploration. Between the years 1789 and 1795, when the Spanish finally left for good, the Moachat may well have been the most thoroughly studied indigenous people of their time.

The Nootka Convention of 1790 defused the crisis between Britain and Spain. Quadra met with Captain George Vancouver in August of 1792 to settle some of the details and a final agreement was signed on January 11, 1794, in Madrid. The Spanish abandoned their establishment in Nootka Sound on April 2, 1795.

Already sea otters were becoming scarce along the outer coast of Vancouver Island. Fur traders now focused their efforts farther north, along the mainland coast and in the Queen Charlotte Islands. Fewer ships called at Nootka Sound every year. Often as not, those that came stopped for wood and water rather than furs. After Maquinna's attack on the trading ship *Boston* in 1803, the traffic slowed to a trickle. Around 1825 the Hudson's Bay Company began to establish itself on the mainland coast and the maritime fur trade more or less ceased to be. By then, Nootka Sound, once the chief and most famous port of call on the northwest coast, had long since slipped back into obscurity. And the rest, as they say, is history.

The great challenge in trying to grasp the intricate local history of west coast Vancouver Island — or any other place — is to maintain an accurate sense of proportion, a precise sense of the temporal relationship between disparate events.

We suffer from a foreshortened perspective regarding the past. Recent events loom large, obscuring our view of earlier times. Entire eras shrink and diminish, taking on an air of insubstantiality as they recede into the past until whole laborious lifetimes can be dismissed in a word or two. The past narrows along chronological lines of perspective, reaches the mental vanishing point and disappears from view.

Our perspective is also distorted by a tendency to dwell over-much on crisis events while passing too quickly over more ordinary stretches of time — there is a marked *rubato* to our recollection of history. People here talk about the arrival of European explorers as if that set of events happened yesterday, more or less. But, in fact, more than two hundred years have elapsed since those first encounters between MamaⱵni and Moachat. That's a lot of water under the bridge: three lifetimes or more, ten generations. It's almost a hundred and fifty years since the first traders came to Stubbs Island. It's over a hundred years since immigrants started to settle the district of Tofino. As a tribe, the MamaⱵni are no longer newcomers.

So we need some means of keeping stricter time — of giving each era its proper weight. Suppose we were to cram the recorded history of west coast Vancouver Island into precisely 60 minutes. One hour by the clock, or better yet, one hour by the stopwatch — an old-fashioned model with the minute hand sweeping around the face.

So here we go:

Don Juan Pérez stops at Nootka Sound, August 1774, trades briefly with the Moachat, then lifts anchor and flees into the gathering gloom. *I start the watch*. Elsewhere: Frederick the Great rules Prussia, Louis XVI is on the throne of France and Catherine the Great is Empress of Russia. George III rules England between bouts of madness, the Boston Tea Party has just taken place and America is poised on the brink of revolution.

One minute into the hour: Captain James Cook arrives with his squadron *Resolution* and *Discovery*, March 1778. Cook's and Ledyard's accounts are published and the extraordinary value of sea otter pelts becomes public knowledge. *Three minutes*, August 1785, Captain John Hanna arrives, the first in a host of maritime fur traders. Event follows event, fast and furious. *Four minutes,* 1789, the Spanish under Don Estéban Jose Martínez establish themselves at Nootka Sound. The Pérouse and Malaspina expeditions arrive and depart. Eliza, Alberni and Mozino come and go. Captains Quadra and Vancouver agree to disagree. *Five minutes*, the controversy between Britain and Spain is resolved and the Spanish evacuate their stronghold. The trade in sea otter pelts peaks and begins to decline. The traffic of fur-trading vessels slows to a trickle, especially after the Moachat under Maquinna attack the *Boston*, 1803, and the Tla-o-qui-aht under Wickaninnish (allegedly) attack the *Tonquin*, 1811.

By then, the Americans have declared their independence (1776), the British have surrendered at Yorktown (1781) and the U.S. constitution has been ratified (1787). The French have had their revolution too (1789). Napoleon becomes Emperor (1804), retreats from Moscow (1812) and meets his Waterloo (1815).

Thirteen and a half minutes, the Hudson's Bay Company begins to establish itself on the coast of what is now British Columbia, starting with Fort Langley on the lower Fraser River, 1827. The maritime fur trade more or less ceases to be. *Eighteen minutes*, March 1843, chief Hudson's Bay Company factor James Douglas establishes Fort Victoria; Native people along the west coast of Vancouver Island start carrying their trade goods south. Vancouver Island becomes a British crown colony, 1849, with Governor Douglas trying to balance his two roles: fur trader and head of the new colony. The first immigrant settlers make their appearance.

Elsewhere: Victoria succeeds to the British throne (1837). Upper and Lower Canada are united (1840). The boundary of the Oregon Territory is settled (1846). Civil war leaves Switzerland a federal state; Karl Marx and Friedrich Engels publish the *Communist Manifesto*; gold is discovered at Sutter's Mill in California (all 1848).

Twenty-one minutes, 1855, Banfield and Frances establish the first seasonal trading post in Clayoquot Sound, at Stubbs island. The first gold rush on the Fraser River, summer of 1858, spells an end to the fur-trading era and the Hudson's Bay Company hegemony. The Crown Colony of British Columbia is established. Douglas is appointed governor of the twin colonies, trying to reconcile a serious conflict of interest between First Nations and the tidal wave of prospectors and settlers. At about the same time a minor gold rush to the Bedwell River brings prospectors into Clayoquot Sound. Fur sealing begins and the first Native hunters are hired to go out on the sealing schooners. The west coast of Vancouver Island is surveyed. Native peoples suffer the first serious outbreaks of smallpox.

Elsewhere: Tsar Alexander II abolishes serfdom in Russia after the Crimean War (1854–56). Darwin publishes *On the Origin of Species* (1859). Italy is united (1860). The United States of America plunges into Civil War (1861–65). Bismarck becomes chief minister in Prussia (1862).

Twenty-four minutes, 1864, Douglas retires as governor. Joseph Trutch, new Commissioner of Lands and Works for British Columbia, reduces many previously established Indian Reserves, while amassing a personal fortune. August 1869, a sawmill is established at the head of Alberni Inlet under the management of Gilbert Malcom Sproat. The united colonies of Vancouver Island and British Columbia join Canadian Confederation, 1871. The Reverend A. J. Brabant, first missionary on the west coast of Vancouver Island, establishes himself at Hesquiat in 1874. That same year, Frederick Christian Thornburg becomes the first permanent trader in Clayoquot Sound with a post on Stubbs Island.

Elsewhere: slavery is abolished in the United States (1863–65). The Dominion of Canada is established (1867). Germany is unified under Kaiser William I (1871). The *Challenger* Expedition takes to the sea (1872–76). Alexander Graham Bell patents the telephone (1876) and Thomas Edison the phonograph (1877).

Twenty-seven minutes, the 1876 Provincial-Federal Commission on Indian Reserves, under the leadership of Gilbert Malcom Sproat, makes the last real attempt to deal fairly with Native claims. Not one of the reserves recommended by the Commission is actually set aside by the province. In 1880, Sproat is forced out, replaced by Peter O'Reilly, Joseph Trutch's brother-in-law. In the late 1880s, O'Reilly defines Native reserves along the west coast of Vancouver Island. First Nations become a minority in the new province, though not on west coast Vancouver Island. Even so, the aboriginal way of life here changes profoundly through the second half of the century. Men and women from west coast communities travel far from home on sealing schooners, to Puget Sound hop fields, to new salmon canneries in Victoria and the Fraser Valley. They return to their villages with wealth and stories, but also smallpox, measles and whooping cough. Christie Residential School for Native children is established in 1899 at Kakawis.

Elsewhere: Battle of the Little Bighorn, aka Custer's last stand (1876). Joseph Swan invents the light bulb (1878). Karl Benz sells his first motorcar and the Canadian Pacific Railway is completed (1885). The Industrial Revolution speeds urbanization in Britain (1880s). Massacre at Wounded Knee (1890).

Twenty-nine minutes, the Sutton brothers, William and James, establish their sawmill and store in Ucluelet, 1884. In 1893, John Grice, Tofino's first settler, pre-empts land at the west end of what is now Main Street. The first salmon cannery on the west coast of Vancouver Island opens in 1895 at the mouth of the Kennedy River. By the turn of the century, *thirty-three minutes*, Ucluelet has 15 settlers and the telegraph line is complete to Port Alberni, with extensions planned to Tofino, Clayoquot and Ahousaht. By the end of the first decade of the 20th century, nearly sixty settlers have taken up land behind Long Beach, and construction begins on a road between Tofino and Ucluelet. The Lennard Island light station goes into operation in 1904. In 1905, the first export sawmill on the west coast of Vancouver Island is established at Mosquito Harbour on Meares Island; it is bankrupt by 1907. By 1910 fur seals are on the brink of extinction; sealing is banned by international agreement.

Elsewhere: Marconi invents wireless telegraphy and the brothers Lumière screen the first moving picture (1895). Queen Victoria celebrates her Diamond Jubilee and miners rush to the Klondike (1897–98). The Wright brothers make the first sustained airplane flight at Kitty Hawk (1903). Albert Einstein publishes his *Special Theory of Relativity* (1905). San Francisco is destroyed by earthquake (1906). Henry Ford sells his first motorcars (1908).

Thirty-six minutes, World War I begins, August 1914, spelling an end to rapid development on the west coast of Vancouver Island. Many homesteads are abandoned. But fishing becomes, for the first time, a mainstay of the local economy, an important source of seasonal employment. In Clayoquot Sound, the industry's main strength is a newly established community of Japanese-Canadian fishermen and their families.

Elsewhere: Einstein publishes his *General Theory of Relativity* (1915). In Russia, the Bolsheviks seize power (1917). In the United States, Prohibition begins and women get the vote (1919).

Growth is slow after the war, but in 1925, *forty minutes*, large schools of pilchard start turning up in west coast inlets. By 1927, there are 26 reduction plants from Barkley Sound to Kyuquot. Overnight the pilchard fishery becomes the region's major industry and, just as abruptly, it vanishes. October 1929, the stock market crash brings the fishing boom to an end. The pilchard stop coming into the inlets. As seasonal work disappears, west coast Vancouver Island families become dependent on prospecting, trapping, government work, anything they can find. A rich seam of gold is discovered in Zeballos, the one bright point in an otherwise bleak picture.

Elsewhere: things are not much better. The Great Depression follows the crash of '29. The Statute of Westminster (1931) gives powers of self-government to the Dominion of Canada. Franklin D. Roosevelt is President of the United States (1932); he launches the New Deal and ends Prohibition. Adolf Hitler is appointed Chancellor in Germany (1933).

On the west coast, the Depression lasts until the outbreak of World War II, September 1939, *forty-three minutes*. In 1942, Japanese-Canadian families are "evacuated" from the west coast and their goods are seized. The war sees the first large-scale industrial logging on the west coast of Vancouver Island under the cost-plus Spruce Account. Elaborate coastal defences are constructed along with a major airport at Tofino, a sea-plane base in Ucluelet and, at long last, a road between the two communities.

Forty-five minutes, World War II is over, 1945. The industrial development of west coast Vancouver Island begins in earnest. Forest industry giants MacMillan Bloedel and British Columbia Forest Products are granted tree-farm licences that give them long-term control over enormous swathes of forest; large tracts are clearcut. A road is opened, finally, between Port Alberni and the west coast,

September 1959. The Brynnor iron-ore mine, a major operation, is established at Maggie Lake and Toquart Bay, 1962. The fishing industry recovers spectacularly from its war-time slump.

Elsewhere: the first hydrogen bomb is tested (1952). The Treaty of Rome creates a European Economic Community and the USSR launches the first artificial Earth satellite, *Sputnik* (1957). John F. Kennedy is elected President of the United States (1960) and is assassinated three years later. Yuri Gagarin becomes the first human being in space and birth control pills are marketed for the first time (1961). The first U.S. troops go to Vietnam (1964).

Fifty minutes, the Brynnor Mine closes, 1966, a scant four years after commencing operations. The 1960s and 1970s see a growing resistance to unfettered development. There is public pressure to save the most beautiful areas along the west coast. Pacific Rim National Park is established, 1971. The environmental organization Friends of Clayoquot Sound is established, 1979.

Elsewhere: Neil Armstrong becomes the first human being to set foot on the Moon (1969). The last U.S. troops leave Vietnam; the world suffers through the OPEC oil crisis (both 1973). Margaret Thatcher becomes the first woman Prime Minister of Great Britain (1979).

The 1980s and 1990s are characterized by a crescendo of environmental protest. In 1984 and 1985, *fifty-five minutes*, a blockade at Heelboom Bay prevents MacMillan Bloedel from logging Meares Island long enough for the Tla-o-qui-aht and Ahousaht First Nations to secure an injunction against logging, pending the outcome of their treaty negotiations. Protestors face loggers and police elsewhere: Sulphur Passage in 1988, Bulson River in 1991, Kennedy River in both 1992 and 1993 — almost nine hundred people are arrested in a peaceful protest against ongoing logging in Clayoquot Sound. The provincially appointed Scientific Panel for Sustainable Forest Practices in Clayoquot Sound publishes its final report in 1995; all 139 recommendations are endorsed by the B.C. government.

Fifty-seven and a half minutes, the Province of British Columbia enters into an Interim Measures Agreement with the central region Nuu-chah-nulth chiefs, 1994, the first step toward negotiating a treaty. Under the agreement the provincial government and First Nations will share responsibility for coordinating land-use decisions in Clayoquot Sound, pending the outcome of treaty negotiations. MacMillan Bloedel closes its Kennedy Lake Division and enters into a partnership with the Nuu-chah-nulth to establish Iisaak Forest Products, a joint venture that will take control of the corporation's tree-farm licence in Clayoquot Sound. The fishing industry collapses; trollers are particularly hard hit. The same period sees rapid expansion of salmon farming and the tourism and hospitality industries. In 2000, *less than one minute left on the clock*, Clayoquot Sound is designated a UNESCO Biosphere Reserve.

Behind that litany of crises, ordinary life goes on. And there lies the other challenge in trying to make sense of history. Extraordinary events — by definition unrepresentative of their times — loom out of all proportion. The lives of a few key actors hog our attention, obscuring the supporting cast. It is hard to maintain a proper appreciation of depth, texture and complexity. Rarely is there a record of routine daily events. You have to rely on your imagination.

> *Who built the seven gates of Thebes?*
> *The books are filled with names of kings,*
> *Was it kings who hauled the craggy blocks of stone?*
> — Bertolt Brecht, *A Worker Reads History*

We bank left over Friendly Cove and cross the mouth of the inlet, heading back toward Tofino. I'm still looking for traces of the relict landscape, even though it has some obvious shortcomings as a means of raising the dead. For one thing, like a photograph, it tends to fade pretty quickly. Think of all the work, the sweat, the anxiety expended here over the years. All that sawing of boards and hammering of nails and washing of linen and fish caught and meals cooked and goods traded. Wearisome labour, year after year, for lifetimes. Huge effort and surprisingly little to show for it. Nature is an efficient undertaker out here.

There is, for instance, almost nothing in the relict landscape of Clayoquot Sound to suggest the early days of MamaⱢni exploration; that era survives only in a handful of place names: Tofino Inlet, Flores Island, Estevan Point. Tofino Inlet was named in 1792 by a party of Spanish explorers under Dionisio Alcalá Galiano and Cayetano Valdés, to honour their teacher, the prominent hydrographer, Rear Admiral Vicente Tofino. Flores Island is named after Don Manuel Antonio Flórez, Viceroy of New Spain. Estevan Point is named after the first commander of the Spanish settlement at Nootka.

The maritime fur trade is only slightly better represented. There are place names — Meares Island, Mount Colnett, Wickaninnish Bay — and also one genuine fragment of relict landscape, traces of Fort Defiance, the base that American fur trader Robert Gray established in the fall of 1791 at Adventure Cove in Lemmens Inlet. Captain Gray and his associate Captain John Kendrick used Clayoquot Sound as a base, thereby avoiding the heated politics of Nootka Sound. In the winter of 1791–92, Gray's party built and launched the sloop *Adventure* at Fort Defiance. Their relationship with the Tla-o-qui-aht and Chief Wickaninnish, begun well with much trading and visiting in the fall, soured dramatically the following spring. On February 18, 1792, Gray's party foiled (as they thought) a surprise attack by the Tla-o-qui-aht. The fort was abandoned on March 25, 1792. Two days later, Gray had the village of Opitsat, apparently uninhabited at the time, destroyed:

> *March 27. I am sorry to be under the necessity of remark-ing that this day I was* sent, *with three boats all well*

> *man'd and arm'd to destroy the village of Opitsatah. It*
> *was a Command I was no ways tenacious of, and am*
> *grieved to think Capt. Gray shou'd let his passions go so*
> *far.* This *village was about half a mile in diameter, and*
> *contained upwards of 200 Houses, generally well built for*
> Indians; *every door that you enter'd was in resemblance*
> *to an human and Beasts head, the passage being through*
> *the mouth, besides which there was much more rude*
> *carved work about the dwellings some of which was by no*
> *means* inelegant. *This fine village, the work of Ages, was*
> *in a short time totally destroy'd.*
>
> — John Boit, Fifth Mate, Log of the *Columbia*

Gray and Kendrick weren't the only traders to do business here. Like Maquinna to the north, Wickaninnish was one of the Sea Otter Chiefs, a wealthy and powerful middleman who controlled trade with the Mamałni, preventing lesser tribes from dealing directly with the newcomers. During the last decades of the 18th century, until the supply of sea otters ran out, the south end of Clayoquot Sound would have been busy with international trade. The two Tla-o-qui-aht village sites, Opitsat and Echachis, would have been the main points of contact between traders and the Natives. But for physical evidence of that trade, nothing remains.

The same is true of the land-based trade. With the decline of the maritime fur trade in the early years of the 19th century, the west coast of Vancouver Island found itself further and further from the beaten track. Fixed trading posts became the new points of exchange. Native people on the west coast of Vancouver Island were forced to carry their goods to the east coast of the island or south to Victoria. Land-based traders came late to Clayoquot Sound. The first seasonal trading post was established on Stubbs Island sometime around 1855 by Banfield and Frances Ltd. With sea otters in short supply, Messrs. Banfield and Frances exchanged their goods for salted fish, miscellaneous furs (mink, martin, a few fur seal) and shark oil (used for lighting, for greasing skidways in logging operations and for the lubrication of gears and bearings). The island's first resident trader, a Dane by the name of Frederick Christian Thornburg, arrived in 1874 to take charge of what he called the Clayoquot Station. But of those activities no trace remains.

The area's first resident missionary, Father A. J. Brabant, also arrived in 1874. Father Brabant's original church, constructed that fall at Hesquiat, and its replacement are both long gone. Even Christie Residential School, established in 1899 at Kakawis on Meares Island and in service to 1972, has been lost to fire. Kakawis, Opitsat, Echachis and other places we could name are relics of a sort, I suppose: they represent the efforts of Peter O'Reilly, Commissioner of Indian Reserves during the late 1880s. On a topographical map, all these places are marked "IR" for "Indian Reserve."

It was not our finest hour. British Columbia had joined the newly established Dominion of Canada in 1871. Under the new arrangement Indian Affairs were nominally a federal responsibility, but control of crown lands fell to the province. The reserves that British Columbia had set aside before confederation were considered inadequate and the province was under pressure from Ottawa to provide a more equable settlement of Native claims. The commission was supposed to be a joint federal–provincial effort. But Commissioner O'Reilly was the provincial government's man and the government was a government of and for settlers. The legitimate needs of aboriginal peoples or even simple justice were not considerations. O'Reilly's real job was to free land from Native claims and open the country for development; the question was not how much to reserve, but how little.

It's not until we get into the 20th century that we start to see any real accumulation of relict landscape: traces of pre-World War I homesteads; pilings for the mill at Mosquito Harbour; foundations at Lennard Island light station (the original tower and light being long gone); remnants of telegraph lines; the gardens at Stubbs Island and

Hesquiat Harbour. There are traces of World War II everywhere: pilings that emerge from the beach every winter, foundation depressions on Stubbs Island to mark the abandoned homes of Japanese-Canadians. And, not least, Tofino airport. Now there's an artifact: in the midst of all this wild country, one of the largest airports in British Columbia. Three 1,500-metre runways built in anticipation of a battle for the north Pacific and now surrounded by Pacific Rim National Park, established in 1971. Losing altitude on the approach, we're actually flying *into* the relict landscape.

For post-war industrial development we have the clearcuts at Hesquiat Harbour, Millar Channel, Cypre Bay, Fortune Channel, Tofino Inlet and so on. Also the fish-processing plants along the water-front, most of them idle now or processing farmed instead of wild salmon. The fish farms are there; also the hotels, restaurants, galleries. Environmental activism has left us a national park and biosphere reserve — though the environmental movement's greatest legacy by far is the absence of relict landscape: Meares Island, Flores Island and much of Clayoquot Sound are still covered in old-growth forest.

And that, it occurs to me, is one more problem with using the relict landscape to raise the dead. The dead are not always well repre-sented by the artifacts they leave behind; there are narrow limits on the kinds of information the relict landscape can give us about the departed. For one thing, it reflects only that part of human enterprise that generates tangible products. Even then — and this is important — the product and the person who made it are clean different things. A nest is not a bird. A bird is more than a maker of nests. The nest says nothing about the living flash of colour in the forest, the dawn singer. In trying to pretend that human life can be represented by a decaying foundation, a clearing in the forest, a few bricks, we get a distorted idea of what human life is all about. Perhaps the bitter truth is that a human life leaves no tangible trace, not really, not even the body it eventually abandons. Life is one of the ephemera, a passing spark, a quickening. A ghost in the machine. At best the artifacts of a relict landscape are echoes of that spark, as a wave on the shore echoes some distant storm, as footprints in the sand echo the living, breathing person who made them.

To reach any further, we must enter again into the province of imagination. Scratch the surface of the past, even just a little, and you discover that there are whole human lives hidden in there, a swarming multitude, mostly anonymous, forgotten. Who speaks for them? You have to look closely to see the individual brushstrokes, the hand of the labourer or artist at work: the man or the woman. And that considers only the serious business of life, the tangible and proper product of their labours. What did they do for fun, for joy, for love, these people? History — like any story — is necessarily selective. It's only possible to discern and name some few tiny fragments of the actuality. But reality, even past reality, is like a Mandelbrot set. The deeper you delve, the more you reveal, endlessly.

As for man, his days are as grass: as a flower of the field so he flourisheth. For the wind passeth over it, and it is gone; and the place thereof shall know it no more.
— Psalm 103 (A Psalm of David), verses 15 and 16

Epilogue

C'is-a-qis Creek
Tofino

EPILOGUE: East of Eden

What is beautiful is a joy for all seasons, a possession for
all eternity.

— Oscar Wilde

First Street Dock, Tofino. There is the odd patch of blue sky still visible through the clouds massing overhead. Sunshine filters through onto the harbour. But it can't last. Curtains of rain eclipse the distant mountains, and the coming storm has already overtaken the far end of Meares Island. The steep slopes fade away, ridge upon ridge, in a complex and pleasing pattern. A smoke of mist wreaths up among the great trees on the near shore and the crying of mew gulls carries across the water.

And so we come to the beginning of another year. My grand tour is complete, surely the time for a little self-congratulation, the satisfaction of achievement. But not yet. I'm still digesting that trip to Nootka Sound. A serious business, a dangerous business, this pursuit of spirits. I have wakened my ghosts and cannot talk them back to sleep. Even in the morning light, they will not let me be.

Somewhere over Hesquiat Harbour, while I was looking at those clearcuts in the mountains, it came to me that the one thing that best characterizes the historic era on the west coast of Vancouver Island is a steady deterioration of the countryside. We started in the late 18th century with pristine environments and small communities of human beings living in a fair degree of harmony with their natural surroundings. We've ended up in the 21st century with whole mountains stripped bare of trees and rivers empty of salmon. It's a tragedy: paradise lost. As far as wild landscapes and wilderness ecosystems are concerned, the Mama‡ni have been nothing but bad news.

This is the other face of Clayoquot Sound, a less-than-perfect paradise, vulnerable and troubled. It wasn't part of that original view from the south end of Long Beach all those years ago, but I have encountered it many times since. Even in this year of seeking beauty the darker side was never far off, waiting around the next corner or over the ridge. There was no getting away from it.

Flying south we passed other clearcuts, a whole series of them: Stewardson Inlet, the Atleo River watershed and Bedingfield Bay, Whitepine Cove, Cypre Bay, Catface Mountain. I felt more and more discouraged. Is this what the future looks like? Vast scars everywhere you look, even in Clayoquot Sound? Sometimes it seems inevitable in an age so infested with need and greed and increasingly efficient implements of destruction.

Then the aircraft banked and there, up ahead, was Meares Island all covered in beautiful forest. Still.

In 1979, MacMillan Bloedel and the British Columbia Ministry of Forests announced that they were planning to clearcut Meares Island. Tofino people were predictably upset; not only is Meares Island part of everybody's view, it's the community's water source. Government and industry officials may have expected some backlash, but they certainly did not anticipate the active counter-campaign that developed. Nowadays such resistance is relatively commonplace but in those times — an era of unfettered resource extraction — little towns were expected to be more obliging.

Here was a dramatic situation, David and Goliath, a little community of committed people struggling to maintain their beautiful surroundings and water supply against the combined authority of a giant multinational company and a powerful government ministry.

The idea that a whole town might have the audacity to resist industry and government in defence of a piece of wilderness, duly scheduled for development, was a radical and disturbing notion for the time. Even today, it has the power to evoke strong feelings.

Due east from the First Street Dock stands Mount Colnett, at the southeastern end of Meares Island. In 1983, when the provincial government finally decided to ignore the opposition and give MacMillan Bloedel the go-ahead to log, smart money would have bet that Mount Colnett, Lone Cone and all the rest would soon be one enormous clearcut. But there it is, forests intact. On the far side of Mount Colnett are C'is-a-qis Creek and Heelboom Bay, where Ahousahts, Mamaɬni and Tla-o-qui-ahts stood firm at the eleventh hour and refused to allow destruction to have its way.

The standoff at Meares Island echoed through the following decades. There were blockades at Sulphur Passage, at the Bulson River, at Clayoquot Arm. There was "Clayoquot Summer," 1993, when thousands of people came from across the country, from all over the world, to protest the ongoing logging — turning out in the cold light of dawn, scrupulously non-violent, morning after morning. Some 860 people of every imaginable background and philosophy were arrested that summer, apprehended in the act of defending beauty. I hold those people to be heroes; they should have collected awards and citations from a grateful nation. Instead, many received harsh sentences for civil contempt of court; their protest was contrary to court orders obtained by the forest industry. Perhaps someday they'll get their medals; these things take time.

And the 1993 protest was by no means the last word. Nothing is settled: Meares Island could still be logged. The turbulent universe deals out endless hard knocks. Environmental activism is like weeding a garden: the job is never done. Efforts to defend the ecological integrity of Clayoquot Sound continue, necessarily, to the present day.

But determined people *can* make a difference, that's the point. We needn't yield to devastation; it is not inevitable. Change may be coming, given the forces at work. But we can shape it to suit our purposes, our own needs and values. Surely this is one thing that distinguishes our species: a capacity to defy the inevitable, to lift ourselves from the path of least resistance.

So I try not to be disheartened. There is cause for hope. Much is gone, but much remains: the Sydney River valley, Pretty Girl Cove, the Megin and Moyeha, Watta Creek, Flores Island, Meares Island, the Clayoquot River valley. The list goes on. Magic and wonderment are alive in the mountains and misty valleys. Distant, secret, wonderful places are out there still, even in this imperfect paradise. Already these forests, these pristine landscapes, have managed to avoid catastrophe longer than anyone might have predicted. With a little luck and some effort on our part, they may continue to defy destruction for who knows how long.

There is honour in defending such beauty, a virtue or merit that transcends this particular place and time. Human beings are not slaves to entropy. We are not bound by the rush to lowest common denominators. We are builders up as well as cutters down; it is our saving grace and future hope.

Besides which, we're pretty much out of beautiful wild country to run to. If we want to take a stand, this is the time and place. It's now or never.

I can hear the storm coming now, as it moves across the harbour toward me, the lightest hiss of distant raindrops on water. The tide has slackened. The wind is gone. All is quiet.

The air is surprisingly warm; the pineapple express has arrived. I draw a deep breath, a great double-lungful. The moist air is full of the smell of wet forest and the pungent, almost acrid odour of the salty sea.

Soon, scattered raindrops spread rippled rings across the smooth liquid surface below me. The tempo picks up rapidly and the sea begins to jump under the deluge, plunging down, bouncing up: a wild confusion of leaping waters. Individual sounds merge into a soft drumming roar.

> *A thing of beauty is a joy for ever:*
> *Its loveliness increases; it will never*
> *Pass into nothingness; but still will keep*
> *A bower quiet for us, and a sleep*
> *Full of sweet dreams, and health, and quiet breathing.*
>
> — John Keats, *Endymion: Book I*

ACKNOWLEDGEMENTS

In any project with such deep roots, so many years in gestation, one naturally comes to owe a debt of gratitude to many, many people for contributions large and small. Something so simple as a few moments of chance conversation, a book recommended at just the right moment, a single line of poetry heard years ago may have borne substantial fruit in these pages.

My great fear, in attempting any sort of a list, is that I must inevitably miss someone. That would never do. You know who you are, all of you. I thank you most sincerely for your warmth and generosity. I hope the book is a satisfactory return on your investment.

There are a few names that do need to be mentioned particularly. For their special encouragement and support: Tena and Doug Pitt-Brooke; Susan Bloom; Mark Hobson; the Langers; Bruce and Jean Robertson. For the use of his lovely images, Adrian Dorst. Scott Steedman, my editor, for his patient and skilful shepherding, and Michelle Benjamin, my publisher, for affording me the opportunity to speak. The copy editor, Jonathan Dore, for a thousand tiny improvements. The rest of the team at Raincoast, for their expert assistance.

I wish also to specially recognize a tremendous debt of gratitude to the people who originally gathered the knowledge and understanding that is such an important part of this book's essential substance. In a work of this sort it is impossible to properly credit original researchers and thinkers, but I want them to know that the contributions they have made to the common fund of knowledge are very much respected and appreciated.

And like Billy Bray I go my way, and my left foot says "Glory," and my right foot says "Amen": in and out of Shadow Creek, upstream and down, exultant, in a daze, dancing, to the twin silver trumpets of praise.

— Annie Dillard, *Pilgrim at Tinker Creek*

ABOUT THE AUTHOR

photo by Wayne Barnes

David Pitt-Brooke practised veterinary medicine for a decade and a half, with digressions into wildlife research, including breeding falcons, collaring caribou and implanting radios in rattlesnakes. From 1987 to 1995, he was an environmental education officer for Parks Canada in Glacier, Mount Revelstoke, Waterton Lakes, and finally Pacific Rim National Park. In 1995 he established his own communications business and has written on a wide range of topics, from grizzly bears to critical path analysis. In June 2002 he received a Canadian Science Writers' Association Award for "Outstanding Contribution to Science Journalism in Canadian Media during 2001." This is his first book.